BUILDING INNOVATION

COMPLEX CONSTRUCTS IN A CHANGING WORLD

DAVID M. GANN

Published by Thomas Telford Publishing, Thomas Telford Ltd, 1 Heron Quay, London E14 4JD.
Tel. +44 (20) 7665 2464; Fax +44 (20) 7537 3631.
www.t-telford.co.uk

Distributors for Thomas Telford books are
USA: ASCE Press, 1801 Alexander Bell Drive, Reston, VA 20191-4400, USA
Japan: Maruzen Co. Ltd, Book Department, 3–10 Nihonbashi 2-chome, Chuo-ku, Tokyo 103
Australia: DA Books and Journals, 648 Whitehorse Road, Mitcham 3132, Victoria

First published 2000

Also available from Thomas Telford Books

A bridge to the future (including CD-ROM) 0 7277 2714 1
The future of international construction 0 7277 2749 4
The decision makers: ethics for engineers 0 7277 2598 X
Infoculture 0 7277 2597 1
IT in construction design 0 7277 2673 0

A catalogue record for this book is available from the British Library

ISBN: 0 7277 2596 3

Typeset by APEK Digital Imaging, Bristol
Printed and bound in Great Britain by MPG Books, Bodmin, Cornwall

For John and Hilary, Anne, Oliver and Dexter

Contents

List of figures

Credits

All illustrations are copyright of the author or out of copyright except the following.

2.1a & b.	Popperfoto/Reuters
2.2b & c.	American Technology Sublime, D Nye, MIT Press
2.3a, b & c.	Building Systems Industrialisation and Architecture, Barry Russell. Copyright John Wiley & Sons Limited. Reproduced with permission
2.3d.	Building the 19th Century, Tom Peters, MIT Press
2.3e.	Architectural Review, London. 'The Turning Point of Building' by Konrad Wachsmann (Reinhold Publ. Corp.)
2.4.	Ove Arup & Partners
2.6.	Le Corbusier: Maison Dom-Ino 1914. Plan FLC 19209 © FLC
2.7.	The Changing Workplace, F Duffy, Architectectural Press
3.3a.	Building Systems Industrialisation and Architecture, Barry Russell. Copyright John Wiley & Sons Limited. Reproduced with permission
3.3b.	Krausskopf-Verlag, Wiesbaden from 'The Turning Point of Building' by Konrad Wachsmann (Reinhold Publ. Corp.)
3.4c.	Bendix is a registered trademark of Philco International Corporation, a division of White Consolidated Industries, Inc. of the US
4.1a.	National Air Traffic Services
4.1d &e.	Tokyo Metropolitan Government
4.3.	Ove Arup & Partners
4.4.	IPD and BZW (Nabarro 1990, p57)
4.5.	IPD and BZW (Nabarro 1990, p57)
4.6.	NEC Corporation
4.8.	Ove Arup & Partners
4.9.	Reproduced with permission of the Royal Institution of Chartered Surveyors
5.1a & b.	Takenaka Corporation
5.7.	Ove Arup & Partners and Ian Lambot Studio
6.3a.	PKL Ltd
6.3b.	Mitsubishi Ltd
6.6.	Obayashi Corporation
6.9.	Frank O'Gehry

List of tables

Acknowledgements

Three ingredients made this book possible. First, the willingness of a large number of people from many firms, government and industry organizations, and universities who spent time discussing their ideas with me about how buildings are produced and used. Second, the sponsorship of a number of research projects without which it would not have been possible to collect and write up much of the data. Third, the support and encouragement of friends and colleagues too numerous to mention here, who so freely exchanged ideas about innovation in the built environment and the management of technology in project-based firms. Their questions and comments in discussions along SPRU (Science and Technology Policy Research, Sussex University) corridors, in seminars and in correspondence from further afield, were directly responsible for stimulating the ideas covered in this book.

I am indebted to Peter Senker for his help and guidance in the early stages of the book; to Andrew Davies for his pithy comments in the later stages; to Chris Freeman, who inspired me to continue studying innovation in buildings and structures; to Carlota Perez, whose comments and encouragement were particularly helpful in finishing Chapter 6; and to Mike Hobday for the title. During the last five years I have been privileged to collaborate with a talented, dedicated and highly professional team of researchers in the Programme on Innovation in the Built Environment at SPRU. I thank them for their ideas, enthusiasm, patience and humour. I owe special thanks to Charlotte Huggett who has provided wonderful support on the many and varied research projects over the years. Her assistance in preparing and proofreading this text has been invaluable. I am grateful to Tim Venables for his help in preparing the images.

I shall be forever indebted to my late friend and collaborator, Steven Groák, who died suddenly in June 1998. His extraordinarily sharp, creative and visionary ways of addressing practical and theoretical problems in design and production and his wit, integrity and passion for excellence in art and engineering continue to inspire.

Preparation for this book involved interviews with many people in construction and related organizations and I would like to thank all those who participated, including clients and building users, contractors, specialist suppliers and equipment manufacturers, professionals in design and engineering, trades unionists, training experts, academics and researchers, and government officials. Over 400 organizations were visited — of these 216 were in the UK, 68 in the rest of Europe, 97 in Japan, 22 in the USA, 15 in Hong Kong and Singapore and 11 in Australia. In addition, information was gathered during site visits to construction projects and at a number of international conferences and symposia.

I have learnt much through close involvement in research and implementation with a number of companies in the UK and I am particularly grateful to colleagues at Ove Arup & Partners, W.S. Atkins and Willmott Dixon who provided opportunities to learn at first hand how to manage innovation in the built environment.

I benefited greatly from involvement in six Department of Trade and Industry Overseas Science and Technology Expert Missions to Continental Europe, the USA, Japan and Singapore. I am particularly grateful to my colleagues from a number of professional backgrounds who joined these Missions and to the hosts in government and industry from whose knowledge I gained enormously. I should especially like to thank Masao Ando and Shigeaki Iwashita for their assistance in my many visits to Japan.

The book is based on the analysis of data collected between 1986 and 1998 in a number of research projects focusing on the management of technology, organizational change, skill requirements and government policy in construction sectors across Europe, North America and Asia. I am most grateful for the support of the projects' sponsors, including: the Building Research Establishment and Leverhulme Trust; the Engineering and Physical Sciences Research Council (EPSRC); the Department of Environment, Transport and the Regions; the Department of Trade and Industry; the Electrical Contractors' Association; the Construction Industry Research and Information Association; the Construction Industry Training Board; the Construction Industry Board; the Joseph Rowntree Foundation and Scottish Homes; the Ove Arup Foundation; the Housing Corporation; and Atias Corporation, Japan. The Organization for Economic Cooperation and Development (OECD) kindly gave permission to reproduce work from a paper prepared for them in 1997.

The EPSRC and Royal Academy of Engineering provided the principal financial support for completion of this book through their sponsorship of my Chair in Innovative Manufacturing, for which I am tremendously grateful.

David Gann

IMI/Royal Academy of Engineering Chair in Innovative Manufacturing and Programme on Innovation in the Built Environment, SPRU, University of Sussex
http://www.sussex.ac.uk/spru/imichair

July 1999

Preface

It is something of a puzzle that innovation and technical change in the construction industry have received so little attention from economists and historians. No doubt, this is partly because construction was often regarded as a 'traditional' industry of very low research intensity and characterized by considerable conservatism and resistance to technical innovation. Such industries have been generally neglected by comparison with the more glamorous and research-intensive industries such as electronics, pharmaceuticals or aerospace. Until recently, most service industries also suffered from this relative neglect and, indeed, they often did not have any R&D activity at all, being almost completely dependent on innovations made by their equipment suppliers.

However, the absence of formal R&D departments or research activities never meant that technical change did not take place or that these industries were not of great interest. Indeed, although innovation in the construction industry received inadequate attention, it was actually the subject of a few extremely interesting studies, notably the two books by Marian Bowley (Bowley 1960, 1966), the work of the Building Research Station on productivity, and Turin's study of the construction process (Turin 1967). Marian Bowley's books concentrated respectively on construction materials and on the construction industry itself and she was concerned to explain why innovation occurred in certain parts of the industry but was retarded in others. This remains a central issue and this book by David Gann contributes a great deal to our understanding of the problem. It does so primarily because of two great virtues: first, its historical approach and second, its systematic approach.

The building industry has for a long time been concerned not just with the primary need for shelter but also with a widening range of buildings for a variety of other public purposes. As Marian Bowley showed, the demand for new types of building has been more important in stimulating technical and organizational innovation than the need to erect better and cheaper buildings to accommodate existing functions. She also showed that innovations in basic materials, originating from outside the building industry and with markets far wider than construction, played an extraordinarily important role in the evolution of the industry. These themes can only be satisfactorily treated on a historical basis.

The availability of rapidly increasing supplies of cheap iron for a huge variety of applications is universally recognized as one of the distinguishing characteristics of the first Industrial Revolution. The transport of coal and iron, first on the canals and later on the railways, made these materials available all over Britain and other countries embarking on industrialization. While the cotton and iron industries were the leading sectors in this vast transformation, the construction industry was also one of the fastest growing, supplying as it did the new factories and mills, the canals and railways and the new houses for the urban population as well as a growing variety of services and utilities for this population. There was a close symbiosis between the uses of iron in the construction of the typical three- and four-storey cotton mills and woollen mills, the uses of iron in the barges and the lock gates of the canals, the 'iron roads' of the railways and the general progress of industry based on machines. Ironbridge in Shropshire is rightly regarded as a symbol of the dawn of the Industrial Revolution. This did not mean of course that older materials, such as timber, bricks and stone, were no longer used. Typically, new or more abundant and cheaper materials do not displace the older ones but only enlarge the range and capabilities of the industry, i.e. they permit innovation.

This was also the case with the abundant supplies of cheap and good quality steel which became available as a result of a succession of process innovations in the second half of the nineteenth century. Not only did Bessemer steel rails vastly improve the efficiency and durability of railways all over the world, numerous other applications of steel transformed the potentialities of the construction industry. Skyscrapers were the most obvious and spectacular manifestation of this new potential, enhanced by the electric lift. The electrical industry itself was a huge consumer of steel for its new structures and changed the design of new houses, offices and factories. Thus, the history of the construction industry is intimately related to the intensive use of new and cheaper materials in each successive technological revolution. Not only has it been one of the main users of these materials but through their applications it has contributed to a whole constellation of innovations affecting both the production and use of these materials. This was particularly evident in the recent wave of innovations affecting synthetic materials such as PVC, polystyrene, Perspex and polyethylene. Consequently, David Gann's historical approach is precisely what is needed to understand the successive stages in the evolution of the industry.

Nowhere is this more apparent than in the most recent and contemporary developments in information and communication technology. Here David Gann's account leaves behind all earlier interpretations and bases itself on his thorough acquaintance with numerous contemporary developments in the worldwide construction industry. His work on the Japanese construction industry is particularly important as it is in Japan that a new pattern of innovation emerged: innovation based on the in-house R&D activities of the construction firms themselves rather than their suppliers. It is here that the systems approach, which is the second major characteristic of David Gann's book, proves its value. The evolution of the construction industry in any particular country can be understood only in relation to the wider social and economic system in which it is embedded. A network of institutions interact with it and help to determine its specific local features. No other book has demonstrated so well this interdependence of institutional and technical change in construction. I commend it most strongly to all those interested in the taxonomy of innovation as well as those concerned with the construction industry.

Chris Freeman

SPRU

May 1999

1. Building innovation

Technological and organizational changes of historic proportions are transforming the production and use of the built environment. The rate and direction of innovation differs from that experienced in the past. This has major implications for the ways in which we use buildings and infrastructures and for traditional methods and locations of construction activities. A number of interrelated forces lie at the heart of this revolution. These include the diffusion of information and communication technology throughout the economy and society, new methods of manufacturing, and growth of the service sector in older industrialized countries. They include general trends towards globalization of markets and production, and new regions of economic growth around the Pacific Rim and in China. Forces of change also emanate from within construction itself, as firms compete in their quest to secure orders and deliver new products and services.

Rapid changes in the economy and society create demands for new construction work and renewal of the built environment. Processes of production, distribution and consumption are changing such that new facilities are needed for the extraction of raw materials, processing, manufacture, retail and service sector activities. New infrastructures are required for transmitting information, transporting people, goods and services, and providing basic utilities such as water, sanitation and power. Rapid urbanization in many parts of the world is resulting in increased demand for housing. While Asia builds anew, Europe and North America repair, refurbish and modernize.

If pressures for change in buildings, structures and infrastructures seem great now, they are unlikely to relent as the forces that shape society continue to ebb and flow. Waves of innovation in information technology, new materials, genetic and biochemical engineering create possibilities for further economic growth and the need to develop new living places and work spaces. The impulse to modernize, creating new goods and services, new methods of production and transportation, opening new markets gives rise to what Schumpeter called the 'perennial gale of creative destruction' (Schumpeter 1976, p. 84). The negative, destructive aspects of development processes are themselves having major consequences for ways in which we produce, use and decommission buildings and structures. Concerns over degradation of the natural environment spur the need for environmental protection, such that

Fig. 1.1. Building anew and restoring the old in a changing environment – The Louvre, Paris (1981–1993)

environmentally clean buildings and structures are increasingly in demand, stimulating innovation in building technologies and methods. These pressures are also changing the location of construction, away from green-field sites towards brown-field areas and urban regeneration. Moreover, the uneven pace of social change and what Perez calls 'mismatches' in the potential to exploit new technologies within existing social and institutional frameworks produce tensions between our experiences and the reality of innovation in the built environment (Perez 1983). The desire and ability to preserve our heritage—enshrined in great buildings from the past—from the onslaught of modernization and development, itself depends upon processes of repair and maintenance that utilize innovative tools, techniques and materials (Fig. 1.1).

The purpose of this book is to explain how broad sweeping technical, economic, social and political changes relate to innovation in design and construction and renewal of the built environment. Such changes are important from an economic perspective within construction and its supply industries, as well as for wider issues of economic prosperity. Conventional, mainstream concepts of innovation are not always applicable for understanding and managing processes of change in the built environment. This book offers new approaches to the management of technology in construction, focusing on the role of project-based firms, with the aim of improving quality, technical capability, reputation and profitability.

In most OECD countries, construction activities contribute around 7 per cent of the total value of goods and services (or gross domestic product (GDP)): the figure is higher if the value of construction-related materials and components is included. In some fast growing, newly industrializing regions (for example, until recently, Japan and Korea) construction accounts for around 12 to 14 per cent of the GDP. These are large industries, often employing millions of people, accounting for 8 to 9 per cent of economy-wide employment in 1994 (OECD 1997). About half the construction output is in markets for residential buildings. The proportion of work carried out as repair and maintenance varies from nearly 50 per cent in the UK to probably no more than 25 per cent in Japan (although accurate data for Japan are unavailable).

Figure 1.2 shows data for value-added OECD construction activities, excluding the value of construction-related materials and components. Taking a snapshot of output in 1970 and 1994, it appears that construction activities have declined in relative terms as a proportion of economy-wide output. Construction output and employment appears to have generally stagnated or declined in OECD countries over the last 30 years, with the exception of Japan, where construction markets remained

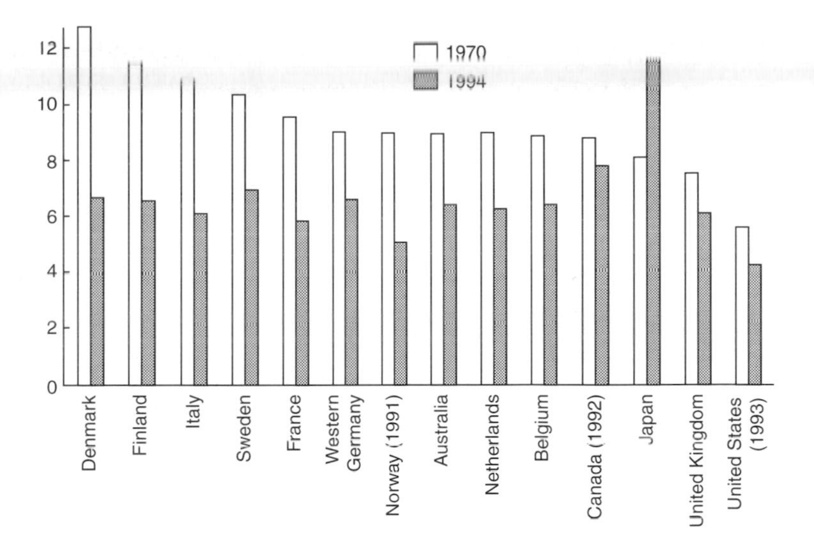

Fig. 1.2. Constructions's share of GDP, 1970 and 1994 (percentages). Note: compound annual growth rates in percentages at constant prices (source: OECD 1998, p. 151)

buoyant until about 1990. Figure 1.3 illustrates the downward trend in construction value-added as a percentage of GDP over the past 20 years from 1977 to 1997, in selected OECD countries. In absolute terms, construction output may not have

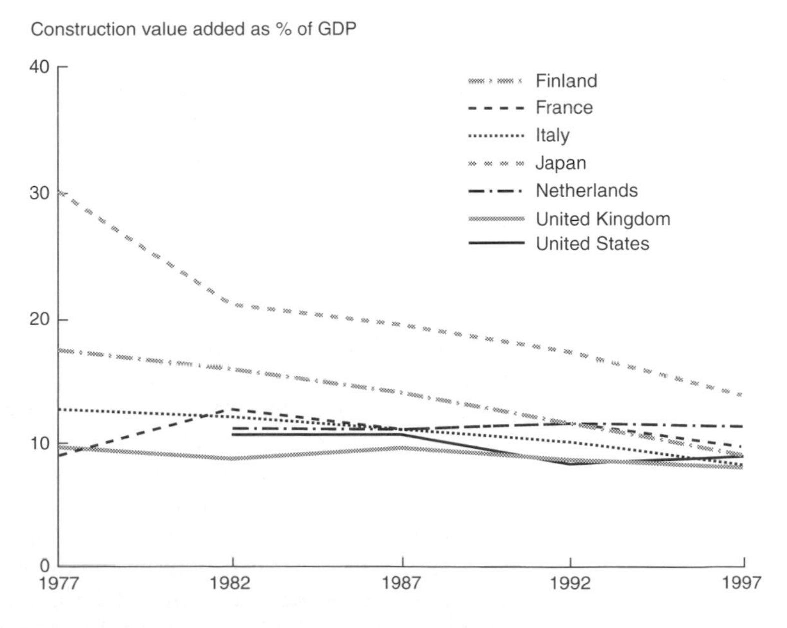

Fig. 1.3. Trends in construction value-added as a percentage of GDP, 1977 to 1997 (source: IMF and OECD data, various years, with thanks to Maurizio Grilli)

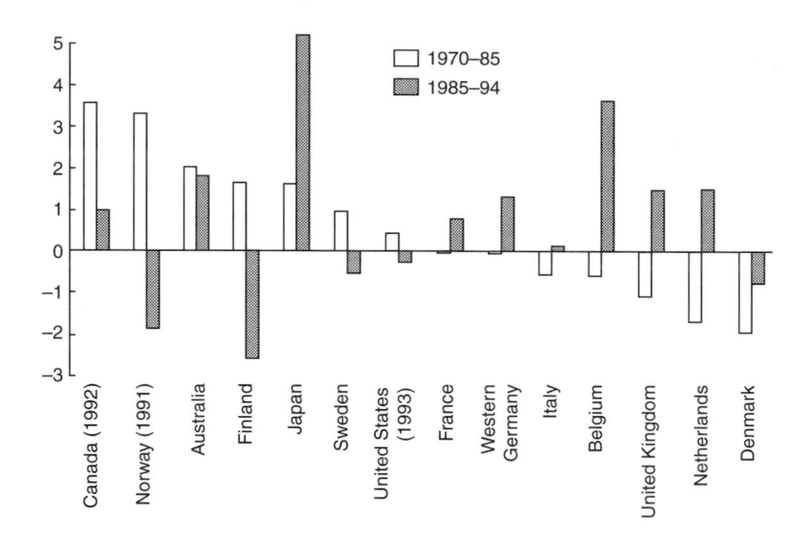

Fig. 1.4. Value-added growth in the construction sector 1970–1985 and 1985–1994. Note: compound annual growth rates in percentages at constant prices (source: OECD 1998, p. 151)

declined much, but as a proportion of GDP it has. Japan has enjoyed higher construction outputs as it has invested wealth in cities and infrastructures since the Second World War. But even here, market growth has gradually slowed as modern infrastructures and buildings have been completed.

Figure 1.4 shows value-added growth in the construction sector, with only limited growth in some countries during the period 1980–1994. Japan is the exception again, where value-added growth rates exceeded 5 per cent in this period.

Markets are segmented according to different product types: housing, commercial and industrial buildings, civil engineering structures and infrastructures, public works, and repair, maintenance and improvement to existing facilities. Demand for buildings and infrastructures fluctuates with business and investment cycles and construction markets are therefore cyclical. In several market segments activities are demand-led, particularly where large facilities and infrastructures are required. Here, production is triggered in response to user needs and, in this sense, projects are demand-derived rather than the result of arm's length market transactions, which typify consumer goods industries.

Government regulatory and procurement policies continue to have a strong influence on demand for construction products and play an important part in shaping the direction of technological change. Regulations on health and safety, land use and rent, planning permits, building standards, energy use and emissions control are necessary to protect consumers and the environment and to create a level playing field for suppliers (Hoj *et al.* 1995).

However, excessive regulation or unnecessarily complicated procedures can contribute to high price levels or restrict customer choice (Gann *et al.* 1998). They may also be used to discriminate against foreign contractors or serve as a barrier to international trade in construction materials and services. For example, within the European single market, slow progress towards harmonizing standards and technical specifications creates a significant barrier to the free circulation of construction products and services (European Commission 1995). Markets are therefore often institutionalized, and regulatory or political pressures may determine the selection of projects and suppliers. Selection mechanisms are usually far more complex than in markets for mass consumption goods, upon which much of the conventional wisdom about supply and demand is based. Demand for large projects is intermittent, with investment taking place over several years. Suppliers for very large projects often require specialist technical know-how, available only from a small number of European, North American or Japanese firms. These firms compete in international markets, contributing to export earnings for their countries of origin.

Construction's significance to wealth creation and quality of life extends far beyond its direct economic contribution. The products and services provided by construction create an infrastructure that supports existing and newly emerging social and economic activities. A well provided, high-quality built environment helps to facilitate wealth creation and improve living standards. If inadequate or inappropriate buildings and structures are produced, or they are poorly maintained and adapted, then social and economic life is compromised. The ability to meet new demands and improve performance through the management of innovation therefore has a direct bearing on the development of construction capabilities and its contribution to economic growth and social wellbeing.

1.1. Ghosts in the system

A number of ghosts haunt construction. Perhaps the most irksome are the difficult relationships between suppliers and between producers and users. This is partly because the built environment has a wide variety of customers, clients and organizations representing users and public interests. Each poses a multitude of different and sometimes conflicting demands that often change over time, with subsequent occupants of buildings requiring quite different facilities and functions from those originally intended. Such conditions result in a complex chain of demands, many of which are not heard at the right time, if at all, by those involved in design, engineering and production decisions. These difficulties in understanding requirements can result in the production of buildings that are inefficient to operate and uncomfortable to occupy, draining users' resources through the need for increased expenditure on repair, maintenance and refurbishment.

The reasons for these problems stem from a number of other ghosts in the system, such as the tendency for design to be separate from construction, and for activities to be passed from one firm to the next with inadequate information. There are other structural deficiencies too. For example, the inability to learn from one project to the next, and poor or inappropriate operative and professional skills. Processes tend to be wasteful and there is often uncertainty over whether projects will be delivered on time and to budget.

Industrial structure varies by country but, in general, the majority of construction and related professional services firms are very small, usually working in local or regional markets. This reflects the history of construction that developed slowly from local and regional activities in which local materials and labour were combined to meet particular market needs, often associated with geological and climatic conditions. In European Union countries, 97 per cent of the 1.8 million construction firms remain very small, employing less than 20 people (Atkins 1994). Many use outdated technology and frequently lack the resources to invest in new equipment and training. Medium and larger firms operate regionally and nationally. Some specialize in particular technical areas and a few are innovative, making use of, and sometimes developing, new technologies. But in many construction markets there is little to differentiate one firm from another except for the price of their bids. This creates market conditions in which fierce price-based competition prevails. Firms focus on managing and passing on risks from one supplier to another, creating an adversarial culture of blame when things go wrong. Only a few companies have the capability and resources to differentiate themselves from their competitors through their reputation for technical excellence. Even fewer are capable of competing successfully in international markets.

Lower rates of productivity growth in construction compared with manufacturing have contributed to a relative increase in construction costs. Output and employment, and therefore productivity in construction, are notoriously difficult to measure. Nevertheless, data suggest that construction has failed to keep pace with performance improvements realized in other sectors. In the period 1970 to 1985, productivity in European construction increased at an average of 0.9 per cent per annum which was low in comparison with other industries (KD/Consultants 1991). Figure 1.5 shows that labour productivity growth rates were modest across OECD countries for the periods 1970–1985 and 1985–1993, averaging around 1.25 per cent per annum for the latter period. Construction in a number of countries,

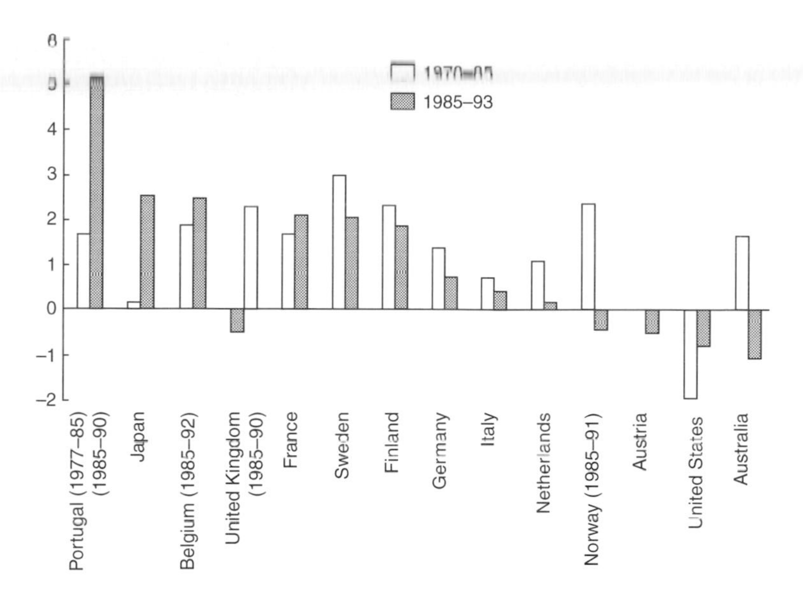

Fig. 1.5. Labour productivity growth in construction, 1970–1985 and 1985–1993. Note: labour productivity is measured by constant price value-added divided by the number employed; the figure shows compound annual growth rates as percentages (source: OECD 1998, p. 153)

including the United States, experienced negative productivity growth. This compares with labour productivity growth of between 3 per cent and 4 per cent annually between 1985 and 1995 in high and medium–high technology industries (OECD 1998, p. 46).

Customers' expectations of construction performance are based not only on perceived improvements in cost, time and quality, but also on comparisons with those of other sectors' products and services. For example, innovation in the computer and telecommunications industries has resulted in rapid and large improvements in performance, driving down the cost of hardware and software. Much slower relative price decreases have been achieved in buildings and infrastructures and there often appears to be little improvement in functionality.

Quality remains unpredictable. This may create pressures to substitute one type of capital good for another, such as information and communication equipment for buildings (Barras 1995, p. 4). Technical progress in information and communication technology is occurring more rapidly than in the buildings and structures where such systems are installed. Barras argues that price and performance differentials across capital goods industries are bringing a new phase of economic development which offers unprecedented opportunities for capital saving through substitution as well as through technical

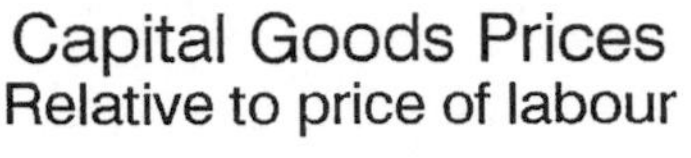

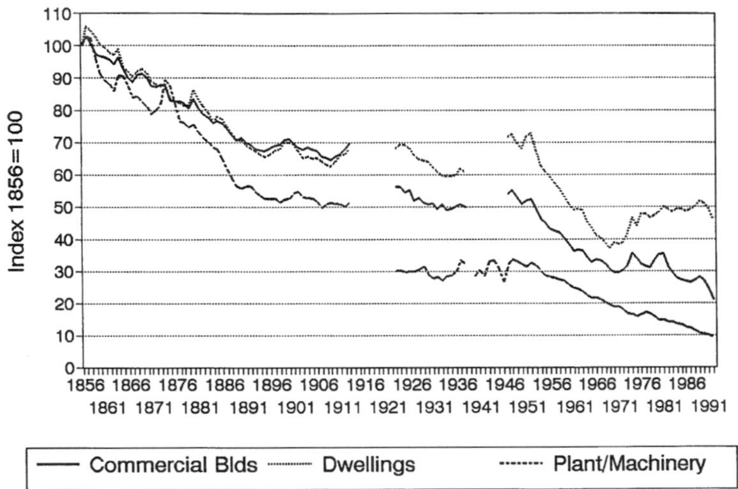

Fig. 1.6. Long-term trends in capital goods prices: commercial buildings, dwellings, plant and machinery (source: Barras 1995)

progress. Figure 1.6 illustrates the long-term trend in relative price differences of commercial buildings, dwellings, plant and equipment. This has consequences for the production and use of the built environment, not least because it relates to new choices in the organization and physical form of cities, requiring less capital invested in buildings and structures, with more going into information networks and equipment. New suppliers are emerging as traditional bricks and mortar businesses are being augmented by firms engaged in weaving broadband digital networks into the built environment, thus changing the nature of buildings themselves. The skills of software and hardware engineers lie at the heart of the digital revolution. The new superhighways of thc latc 20th century have been constructed by electronics and telecommunications engineers. The role of civil engineers, architects and builders as leaders of the process is being usurped. Bill Gates, Steve Jobs and Larry Ellison are the modern equivalents of Brunel, Paxton and Thomas Cubitt. The result: less work for traditional construction firms and more for information equipment and systems industries. This trend is indicated in Fig. 1.7, which shows that the proportion of plant and machinery has increased in terms of gross capital stock relative to commercial buildings and dwellings over the long term. The value of plant and equipment installed within buildings has also increased relative to that of the structure and fabric of the building itself. These trends are consistent with the

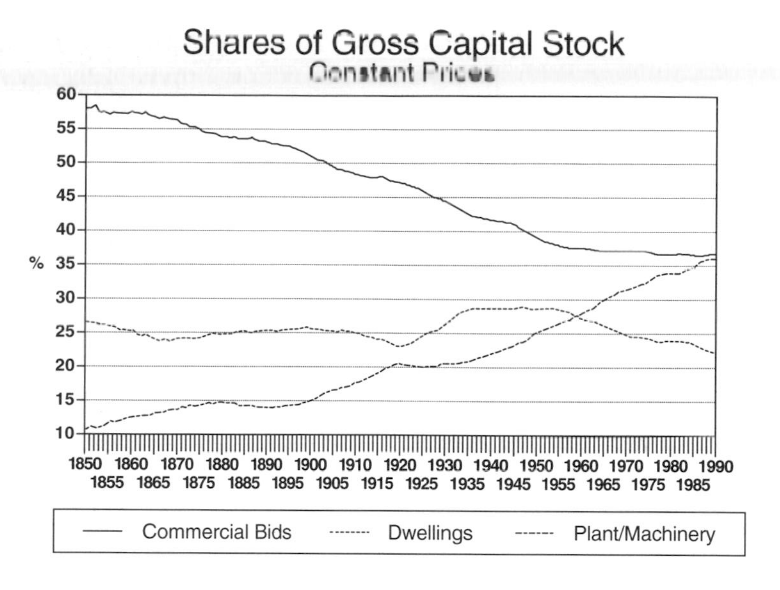

Fig. 1.7. Long-term trends in the share of gross capital stock (source: Barras 1995)

gradual decline in construction value-added as a proportion of GDP over the past 25 years (Fig. 1.2).

Recent enquiries into the performance of construction in the UK and USA provide evidence of construction's shortcomings in terms of cost, timeliness and quality, from a customer perspective (Latham 1994, Wright *et al.* 1995, Bernstein and Lemer 1996, Construction Taskforce 1998). These features are among the reasons why customers are demanding lower construction prices, but this in turn may lead to long-term profitability problems for construction firms, unless they innovate. However, poor profitability itself hinders investment in new techniques, resulting in competition for work based on price alone, putting further pressure on firms' profit margins. When this behaviour persists, a downward price-cutting spiral ensues. Barriers to entry are also lowered because competition tends to be undifferentiated by technical capability. Yet there is evidence that major clients from the utilities, banking, retail and transport sectors expect rates of improvement in construction performance and value for money that outstrip what the traditional sector can deliver. Because of these conditions there may be scope for competition from firms that are able to innovate and deploy new technology, offering more efficient design, engineering, construction and supply.

The relatively weak performance of construction firms is reflected in a number of other indicators. For example, construction has a worse health and safety record than most sectors, and the share prices both of construction and building materials firms underperformed those of many other industries during the 1990s. These aspects of the economics of construction, together with stagnation in output growth, changes in patterns of demand, and the possibility of greater international competition indicate the need to reduce costs and improve quality. Substantial technical change and organizational restructuring is required to achieve this. Yet current construction processes are to a large extent self-optimizing within the constraints of existing skills and rules-of-the-game, fixed in the organizational form of each project. Each player seeks to gain maximum advantage by operating within this framework, and rarely is any one player powerful enough to change the rules such that the production system can optimize at a higher level of overall performance. Innovation, therefore, often takes place within a closed loop on individual projects, where only marginal improvements are experienced overall.

1.2. The need for technological innovation

In this book it is argued that pressures for innovation are strongest when there is demand for radically new types of buildings and structures. This usually occurs during periods of general and widespread technical and economic change. Demand for totally new types of constructed products grew rapidly from the early 1980s onwards, with the need to construct facilities to support activities based on the use of information and communication technology. For example, the design and construction of silicon chip fabrication plants—amongst the cleanest environments on earth—was necessary to produce silicon wafers, one of the key technologies at the heart of the computer revolution. The building of new dealer rooms was necessary to accommodate electronic trading, at the centre of the revolution in financial services in the mid 1980s. Other examples include control and switching centres for computer and telecommunication systems and other infrastructures, including control of air traffic, transportation and utilities systems.

New types of buildings are also needed to accommodate evolving patterns of employment. There has been a shift in employment from manufacturing to services with between 60 and 70 per cent of employees in industrialized countries working in the service sector by the mid 1990s. An increasing number of these work in information processing activities. New buildings, such as call centres, are needed to house these activities: call centres employed around 3 per cent of the European workforce in 1998 (Bateson 1999). Accommodating information and communication technology in the home has also proved to be a major challenge, requiring radical changes in technology, design and housing production processes (Gann *et al.* 1999).

The ability to respond to these challenges is important, not just for construction, but for economic growth and society more generally. The expansion of markets for new information services depends upon the development of a modern communications infrastructure. This includes the design and construction of suitable buildings to accommodate information systems together with new functions associated with their use. Existing buildings, the majority of which are long-lived and were designed before the growth of the information economy, need to be modified. Figure 1.8 illustrates a range of new types of buildings, none of which existed before around 1980.

Other examples of new building products include flexible warehousing and distribution centres, retail and office spaces, together with multi-purpose leisure and events stadia. People are travelling more than ever before, creating demand for new transport facilities. These include large airport complexes to accommodate a wider range of activities, not simply the boarding and disembarking of aircraft and passage through border controls. Expansion of facilities at many international airports is such that they are now of a size, diversity and complexity comparable to small cities, employing tens of

Fig. 1.8. Buildings of the modern world

thousands of people and supporting many millions of users each year. These types of facilities and infrastructures did not exist 20 years ago.

In many construction markets there is increasing demand for flexible, lightweight, multi-function spaces that can be reconfigured for different uses quickly and cheaply. Users prefer engineered solutions that offer greater choice of layout, finish and aesthetic quality. A direct relationship exists between performance in design and construction and efficient and effective operation of these facilities. Comprehension of user–producer relations has never been easy, particularly because of the involvement of so many organizations and interests in convoluted demand and supply chains. Yet rapid shifts in patterns of demand emphasize the need for closer links between producers and users than has hitherto been the case. This is particularly so if user needs are to be properly articulated and fulfilled. In addition, knowledge required for design and construction is expanding because the operation of new facilities often involves their management as part of larger and older technical systems and infrastructures. Thus individual projects must be designed and built within constraints defined by existing systems and the legacies of the technologies they embody (David 1985). Products whose operating requirements and long-term costs are considered in the design process and are planned and built to cope with change may therefore offer greater utility.

Pressures on design and construction companies to innovate and develop new products are likely to continue in the future, with the need to produce facilities for creating and working with biochemical, pharmaceutical and genetically modified substances, as well as new nanotechnology materials produced through manipulation at the molecular scale. Failure on the part of design, engineering, construction and related supply industries to develop these new products could constrain economic investment and growth, diminishing potential benefits accruing from a modern infrastructure.

New product development may be the principal spur to innovation during particular periods of widespread technical, economic and social transformation, but it is not the only driving force of change. Others include new ways of financing projects; the need to comply with evolving international, regional and local regulatory frameworks—particularly concerning health and safety, sustainability and the environment; and issues concerning social exclusion, inclusion and acceptability. Moreover, construction processes themselves have internal dynamics that drive change, as well as those that constrain and retard performance improvements.

Construction is often thought of and excused as a traditional, mature industry that is slow to change. But as Steven Groák argued, it is wrong to see construction as a failed form of manufacturing. It has its own dynamics of industrial development and these are explained in the next chapter. Efficiency, responsiveness and capability to innovate depend to a large extent on the structure of firms, the types of skills employed, management capabilities and relationships with other firms with relevant technical expertise.

Not all construction organizations are passive recipients of changes emanating from other sectors. Whilst many innovations in materials and components are made prior to their installation in the construction process, construction firms nevertheless function as systems integrators and intermediaries in the transformation of technologies from their point of origin to end users. They can play an important part in modifying and developing new technologies, conveying vital feedback from upstream producers to downstream clients and eventual users, and vice versa. The organization of these processes differs according to the types of projects and firms' business strategies. Different ways of organizing production processes in turn create dynamics within the construction system itself, resulting in incentives as well as inhibitors to innovation. In this project-based environment, the capability for innovation in the ways in which projects are organized is often as important as that of managing new technologies.

Forces for technical change are particularly strong among materials and components manufacturers who are often able to invest in long-term research and product development. Many major technological changes aimed at improving construction processes take place away from construction sites and seek to reduce skill requirements on-site. Value-added in construction is increasingly being produced upstream in the supply chain, by component manufacturers. Customers of large projects may also feed the innovation process by funding and participating in research and development activities.

Construction firms have always displayed a peculiar capability for innovation. The site-based nature of production, increasing numbers of different specialisms, relative uniqueness and the changing use of final products and variety of production processes, constantly throw up problems which firms have to solve in a variety of ways. But in spite of an innate ability to deal with change, construction is not generally viewed as an innovative sector, either by those working within it or by outside observers. Innovation prompted by the need to solve problems, often needlessly created elsewhere in the production process, generally lacks direction and can result in further problems which others have to rectify. Hence the paradox of change: construction has internal dynamics that generate innovation, but not necessarily of the type that leads to lasting improvements in performance. What is required is systematic, effective innovation leading to recognizable progress in terms of cost and time reductions in processes and improved quality and functionality of products. In order to achieve this the industries involved need to develop a commitment to learning from previous experience, gaining feedback from users. They need new techniques for measuring improvements enabling assessments of performance over time to be made.

1.3. Project-based, service-enhanced firms

In this book, construction is viewed as a *process* rather than as an *industry*. It includes designing, maintaining and adapting the built environment, and it involves many organizations from a range of industrial sectors, temporarily working together on project-specific tasks. These functions involve planning and design, engineering, supply and integration, erection and installation of a diverse array of materials, components and increasingly complex technical systems. The project-based nature of the work implies that firms have to manage networks with complex interfaces. Correct delivery of products and services requires collaboration between firms. Performance and competitiveness depend not solely on the single firm but on the efficient functioning of the entire network. A technology support infrastructure, including professional institutions, industry organizations and associations, together with mobility of personnel, aids learning between companies and projects. Firms' technology strategies therefore need to extend beyond their immediate boundaries if technologies are to be managed effectively. This raises questions about the management of technical know-how in projects and the management of business and inter-organizational processes within and between firms (Lorenzoni and Baden-Fuller 1995). Many of these issues are similar to those encountered in other project-based sectors that produce complex products and systems, such as aircraft, ships, oil platforms and high-speed trains (Hobday 1998).

Design, engineering and construction firms therefore work in dynamic, multi-technology environments in which they need to manage innovation and uncertainty across organizational boundaries, and within networks of suppliers, customers and regulatory bodies. Knowledge is differentiated and distributed throughout these supply networks. The management of technical know-how has become a significant strategic consideration for suppliers and operators. There is a need for integrity of information between firms, up and down supply networks. Yet many firms tend to manage risk by retaining information crucial to systems integration within their own sphere of control, rather than by transferring know-how between the temporary coalitions of firms with whom they collaborate. At the same time, there appears to be a growing need for suppliers to provide more than basic physical products. Value-added services to support components and systems are needed to assist owners and users in operating, maintaining and adapting buildings and structures. The growing demand for packaged products and service delivery is blurring the traditional boundaries between manufacturing, design, construction and service sectors (Marceau 1997, Gann and Salter 1998, Lester 1998). New forms of production are emerging which centre on *project-based and service-enhanced activities*. Services include financial deal structuring, planning and design, specialist consultancy, customer support and training, supply chain coordination, production and risk management, together with the management of coalitions of interests concerned with project operation, use and facilitics management—including legal, environmental and regulatory governance authorities.

Project-based firms trade on their reputations, based on performance on previous projects. If problems occur on highly visible projects the reputations of such firms may be tarnished, causing difficulties in winning new orders. By contrast, the successful completion of projects may help firms to advertise their capabilities, often assisting them in securing new work. This may also enable the development of wider technical expertise, creating a positive internal culture and making it easier to retain skilled staff. One of the goals of project-based firms is to develop a critical mass of technical capability that drives the business internally like a flywheel, generating new competencies and reinforcing a virtuous cycle of success and growth. For example, Ove Arup & Partners has become a world leader in structural engineering—a reputation consolidated with the completion of the Sydney Opera House (1957–1973). The ability to leverage reputation to win new orders and thus gain further experience in the deployment of technical expertise is of crucial importance to long-term competitiveness. Ove Arup & Partners subsequently provided engineering and acoustical consultancy services in the construction of other concert halls, including the Glyndebourne Opera House (1988–1994). Developing and maintaining a reputation from one project to the next therefore has a direct impact on the ability to win new orders. This in turn affects turnover, profitability and the potential to invest in new generations of technology.

1.4. Approach and structure of the book

How is the production of the built environment changing, and what are the best ways of managing innovation in project-based firms? These are the two central issues at the heart of this book. The results of more than 13 years' research are presented, providing evidence based on statistical data and interviews in over 400 firms, projects, industry organizations and government institutions in Europe, North America and Asia. This material is analysed using mainstream innovation literature, which has hitherto not been applied to the study of innovation in project-based industries or the built environment. The subject matter is broad, reflecting a wide range of product and process innovations occurring over time in the construction and renewal of the built environment.

A few simple questions are raised throughout this work.

- What are the main driving forces of change and their consequences?
- What constraints to improving performance exist?
- What can be done to improve the management of innovation in the built environment?

A route to answering the above was followed by asking a number of subsidiary questions, in an attempt to present a clear, simple framework without losing the contradictions and rich diversity found in the subject matter. The following are examples.

- What role have clients and users played in demanding new types of buildings and structures, triggering innovative approaches to construction?
- Have construction technologies developed historically along a similar path and what are the consequences of the introduction of radically different products or materials?
- How important is organizational change and industrial structure to the particular ways in which construction technologies are developed and used?
- How do skills and learning processes relate to technical and organizational changes on- and off-sites?
- What can be learnt about the management of innovation in construction from ideas practised in other industries?

There have been few attempts to answer these questions in a systematic fashion although there is a useful and important body of work on the sector's economic development, its history and its labour processes (for example: Ball, 1988; Clarke, 1992; Giedion, 1967; Hillebrandt, 1984; Russell, 1981). An approach was therefore needed to explain incremental improvements in existing products and processes, as well as major technological breakthroughs and social, economic and natural events that alter the course of innovation in the built environment. Each historical period is associated with its own particular processes of urbanization and forms of built environment (Mumford 1961). Dual pressures are at work, changing both the *functions* that define the use of buildings and the physical *forms* needed to accommodate them. Distinct patterns of urban development evolve in parallel with each new technological and economic regime and its associated institutional structures (Barras 1995).

The last 200 years have been punctuated by a number of major technical, economic and social changes that have had profound effects upon the built environment (Freeman 1978, Dosi *et al.* 1988, Freeman 1994). For instance, the growth of industrial activities based on the use of steam engines is linked to the emergence of industrial cities in the nineteenth century. The spread of electric power and automobiles is closely related to the growth of twentieth century suburban cities. Widespread diffusion of information and communication technologies is giving rise to what Castells calls the 'informational city' (Castells 1989).

This book divides the period considered into two distinct phases: the machine age, from the nineteenth century to the 1960s, and the digital age—from the 1970s onwards. The book is presented in three parts.

Part 1 provides an historical account of innovation in the machine age associated with the evolution of industrialized construction processes during the nineteenth and most of the twentieth century. It documents the decline of craft processes, the rise of mechanization, production driven by economies of scale and widespread use of technical systems. This period of development in construction techniques was closely entwined with demand for new types of buildings and structures, themselves related to the production of new materials such as iron, steel, glass, composites and plastics.

Part 2 explores technologies emerging in new kinds of buildings and structures associated with the growth of the information society and the digital era at the end of the twentieth century. These include so-called 'intelligent buildings' designed to facilitate new types of work and leisure activities based on the use of information and communication technology. Changes in techniques used in the production of the built environment are also closely associated with the utilization of information systems in design and construction processes.

Part 3 presents a framework for understanding innovation in the built environment, together with ideas about the management of innovation in firms. At the level of the firm, technology policy and the strategic management of resources involve issues of how firms develop their core technical competencies within project-based environments. The extent to which technical competencies are specialized and located in different places within and between organizations affects how they can be deployed, ultimately affecting project performance. The book ends with a blueprint for managing innovation in project-based firms, showing how such companies can deliver better value to clients and improve both profitability and their own potential for growth.

PART 1

THE MACHINE AGE

2. Building innovation in the machine age

Over the past 200 years there have been successive efforts to improve the quality of buildings and structures and to industrialize construction through the development and adoption of innovative techniques. Attempts to modernize construction were stimulated by the pursuit of greater efficiency, often associated with the introduction of new technology, and to meet growing demand for new types of constructed products.

The industrial development of construction has been punctuated by two major turning points in which modern technologies and new ways of organizing production emerged to challenge traditional building methods. The first divide occurred during the middle period of the nineteenth century, when new technologies and organizational techniques used in factories began to be deployed in construction, displacing traditional craft methods. This period is described as the 'first machine age' and it is the subject of this chapter (Banham 1960). The second divide is characterized by the transformation from an industrial to a post-industrial society, which became evident at the end of the 1950s. This was the beginning of the 'information age' and what Daniel Bell describes as the shift from a goods-producing to a service economy (Bell 1974, p. 14). Demand grew for completely different types of buildings and competing forms of industrial organization emerged in construction, challenging traditional craft and existing industrialized methods. Innovation in the digital age is discussed in detail in subsequent chapters.

In construction, unlike in other industries such as motor vehicle production, old processes were not completely swept away during these periods of radical change. Traditional and modern processes coexisted side-by-side. Traditional construction processes were typified by the slow evolution of craft techniques associated with incremental, minor innovations based upon informal learning processes. At the same time new processes, associated with radical innovations and the application of structured, formal scientific and technical knowledge began to emerge.

There are considerable differences between the driving forces of innovation and the technologies of construction in the machine age and those in the digital age. In the machine age technical innovation occurred in developments in building structure and fabric and what are called 'building engineering services' (the mechanical and electrical systems installed within buildings to provide environmental comfort and control). Whilst improvements in these areas continued, the emphasis shifted in the digital age to innovation in electronic systems installed in buildings and structures focusing on communications and control. The arguments provided in this chapter indicate the preoccupations of those involved in trying to modernize the production of the built environment. The examples illustrate how their efforts were often intimately linked with changes in demand for, and the use of, constructed products.

2.1. Craft methods

Prior to the nineteenth century, construction drew mainly upon local sources of labour and materials—a combination which resulted in local types and styles of buildings. Resources from further afield began to be used as demand grew for larger buildings and structures, located and concentrated in new urban areas, or close to mineral deposits and ports. Traditional construction techniques evolved slowly from craft traditions rooted in a labour process typified by the relationship between master and apprentice—the artisan system that had changed little for centuries.

Craft skills combined knowledge of materials, an understanding of the tasks to be done, and the manual dexterity of performing them. Craftsmen often made their own tools and skills were handed down from father to son. The basic unit was the family, producing from a workshop. The building owner, his architect or his agent employed each craft directly. The coordination of various stages of construction was carried out by master craftsmen who employed journeymen and managed their own apprentices and labourers. There were numerous categories of work within building processes during the transition from craft production to its organization in building firms. For example, divisions existed between masters and artificers, masters and journeymen, master artisans and artisans, chief artificers and ordinary artificers (Clarke 1992, p. 72). Master craftsmen organized their activities in associations known as guilds, which were linked to specific trades. They and their guilds embodied knowledge associated with the use of particular materials: bricklayers worked with brick, masons with stone, carpenters with timber, and plumbers with lead. The precise reconstruction of Shakespeare's Globe Theatre on the South Bank of the Thames in London in the mid 1990s provides a good illustration of some of the craft skills originally practised in 1599. These include carpentry, thatching and plasterwork (Fig. 2.1).

The guilds acted on behalf of the masters whose roles included those of materials merchants or tradesmen. Some profiteering masters began to employ more people and gained greater control over the means of production. The demise of the guilds began with the split in roles of masters between those

Fig. 2.1. Reconstruction of Shakespeare's Globe Theatre (1996) using traditional crafts and materials

running contracting firms and those continuing in traditional ways. The organization of construction into general building firms and specialist firms evolved in the mid eighteenth century. In London, speculative builders first appeared in the 1750s, when bricklayers and carpenters established businesses called 'master builders'. In 1815 Thomas Cubitt introduced a new way of organizing the labour process. Instead of arranging with craftsmen to carry out particular jobs, he started to employ them, supervised by craft foremen to work on whatever jobs he had for them. The transition in wage forms, from payment for piece-work to time-based remuneration, signalled a complete change in the way production was organized and in how craftsmen and owners related to one another (Clarke 1992, p. 69). This method of organization grew swiftly and by the 1830s there were several large building firms operating in London. Independent journeymen were replaced by skilled workers and masters were replaced by new employers in building firms. Meanwhile, the role of the apprentice ceased to exist in its traditional form.

The building professions emerged in the 1830s: the Institute of British Architects was the first to be formed in 1834. Professionalization received legal status and royal sanction, and the design and engineering professions guarded their independence from contractors by restricting entry and enforcing strict codes of conduct and performance on their members (Ball 1988, pp. 57–63). The division of labour between design and production had emerged.

Know-how about craft techniques used on sites was rarely documented or codified—it was passed on verbally or by demonstrations from one person to the next, and much existed as tacit knowledge. Craft practices resulted in bespoke buildings, each individually built to the owner's requirements. Even where rows of similar houses were constructed there were variations in each due to the craft methods deployed. This production system resulted in product variations similar to those experienced in the early days of automobile manufacture, where at the beginning of the twentieth century craft-produced vehicles often differed significantly from each other. This was due in part to the sequential fitting process which resulted in 'dimensional creep' (Womack *et al.* 1990, p. 22).

2.2. Industrialized techniques

Building work altered little for hundreds of years until the rise of industrial capitalism brought changes in the organization of production, new processes and technologies for construction, and changes in the type, size and styles of buildings and structures. The nineteenth century was a period of rapid change in the built environment. Technical marvels were springing up all over the rapidly industrializing world, epitomized by the engineering feats of British, then German and French civil engineers such as Fairbairn, Cubitt, Telford, Brunel, Stephenson and Eiffel (Peters 1996). In the USA, engineers vied with each other in their attempts to create evermore spectacular, awe-inspiring technological structures such as the Brooklyn Bridge, Erie Canal, Hoover Dam, the skyscrapers and railroads (Nye 1994) (Fig. 2.2).

There are many examples of innovative construction projects from the nineteenth century. They include the Britannia Bridge (1845–1849) spanning mainland Wales to Anglesey, six times further than any other bridge of that time, and the Suez Canal (1859–1868). Richard Turner's Palm House at Kew (1846–1848) and Joseph Paxton's Crystal Palace (1849–1851) illustrate the application of manufacturing techniques in building assembly, together with military approaches to planning, replacing traditional trial-and-error methods found in many on-site construction activities.

Crystal Palace is a landmark in the history of industrialized construction. It typifies the transformation of the construction site from a place where craft skills honed materials, to a place of assembly of standardized, prefabricated, mass-produced components. Joseph Paxton's design was based on a modular approach

Fig. 2.2. Major nineteenth century engineering projects: Britannica, Eads and Brooklyn Bridges

using many interchangeable components. It was complemented by Charles Fox's manufacturing and construction expertise and William Cubitt's supervisory skills. The design included a light steel structure with a weatherproof lightweight skin, or curtain wall. The framework was designed to be used as scaffolding during erection. Mechanized erection techniques such as roof glazing wagons were deployed. A dry assembly method was used which facilitated the rapid erection of floors and the possibility of dismantling internal fixtures and fittings as usage changed during the building's life. The designer, engineers and suppliers all worked together as a team in one organization. Paxton's background as a gardener (he had built several hothouses and was obviously influenced by greenhouse designs) rather than a career in building helped him to think about construction in a different way (Russell 1981, pp. 33 and 46). Fox's knowledge of bowstring trusses, his construction of large-span railway sheds and use of complete iron systems provided the expertise required to transform Paxton's ideas into a set of detailed engineers' working drawings (Peters 1996, p. 226) (Fig. 2.3).

By the middle of the nineteenth century, the traditional path along which construction techniques had developed was already dividing as a new trajectory of industrialized construction techniques emerged alongside more traditional craft methods. This trajectory was associated with the manufacture of an increasing number of materials and components, rather than their manual production on-site. The evolution of construction materials and component manufacturing capabilities depended upon the invention of production equipment, made possible by the development of machine tools. In the early nineteenth century, skilled craftsmen were the dynamic force behind the development of better and more precise machine tools to solve novel production problems which could not be surmounted in other ways. The principle of interchangeable manufacture, developed in the American armaments industry, was of crucial importance because it led to the acceptance of common measurements and measuring devices, which facilitated the production of standard ranges of parts (Rolt 1965, pp. 137–153). The idea of interchangeable manufacture spread quickly through many industries in North America, such as the production of clocks, bicycles, agricultural machinery, typewriters and sewing machines. It was an essential element in the rise of mass-production. However, it was 50 years before interchangeable manufacture, known as the 'American System', became accepted in Britain.

Such developments in manufacturing were closely related to the sequence of stages in the evolution of industrialized construction techniques. The first stage began in the nineteenth

Fig. 2.3. Crystal Palace

century with the manufacture of simple standard parts, rather than their manual production on-site. The second stage, occurring in the early twentieth century, involved the assembly of combinations of parts in factories to form prefabricated components — these were sub-assemblies of buildings, such as door and window sets. Finally, by the 1950s and 1960s, sub-assemblies were combined together to form large elements of buildings. These were prefabricated as component systems, sometimes using assembly-line methods. This was the era of 'systems-building'.

The invention and use of woodworking machines for sawing, planing and jointing became widespread in the US between 1800 and 1840: this was associated with westward population migration within America, which utilized plentiful supplies of timber for housing. Some of these machines originated from earlier British inventions in Royal Navy shipyards, but were soon transformed in a series of incremental innovations. Benefits from innovations in woodworking machinery were supplemented with those in nail-making machines. Towards the end of the nineteenth century, American woodworking machines were re-imported into Britain. By the 1880s, woodworking machines had revolutionized carpentry in the US and by around 1900 they were having similar effects on carpentry trades in Britain. Carpentry work became more specialized with the new technical division of labour. Master carpenters were replaced by framers or carcassing carpenters, first and second fix carpenters and specialized joiners (Rosenberg 1976, Reckman 1979).

In spite of mechanization, evidence suggests that the reduction in real costs due to the introduction of machinery in the nineteenth century was only moderate. The huge expansion in building activity and utilization of machinery between 1891 and 1901 resulted in a fall in real costs of only 5 per cent. This decline was attributed almost entirely to the introduction of machinery into joinery workshops. The reasons for mechanization were therefore associated with economies of increasing returns due to expansion of markets (Clark 1960).

Developments in brickmaking provide another example of attempts to industrialize construction work. Until the early nineteenth century, bricks were usually handmade to rough dimensions on-site, but by the 1820s production had moved into factories. The introduction of wire cutters and dryers, pug-mills for grinding clay and brick-pressing machines and extruders had transformed the skills of the brick industry, dividing brickmaking from bricklaying (Clarke 1992, p. 73).

Thousands of small brick factories sprang up across Britain, producing distinctive bricks from local clays. The transport costs of both the raw materials (clay and coal for the kilns) and of the finished product meant that it was uneconomic for bricks to be produced far from their point of consumption. By the end of the 1890s, a good hand moulder would produce 1200 bricks per day and this would involve handling some 18 tonnes of clay. However, as transport networks developed, brick production became more concentrated and manufacturers began to reap returns from further investment in automation and from economies of scale. In Britain, the number of brickworks dwindled to 1300 by 1950 and to 300 by 1980, as large-scale producers invested in bigger plants (Massey and Meegan 1982). During the 1970s further investment in modern technologies and expansion of plant capacity resulted in new factories. A 1970s wire-cut brick factory typically produced 40 000 bricks per hour, equivalent to 500 bricks per person hour. An old London Brick Company (LBC) brickworks employing 86 people, producing 600 000 bricks a week, was replaced by a works employing 260 people, producing three million bricks a week— three times the labour force producing five times the number of bricks (Architects' Journal 1981).

The ability to innovate in the production of building materials was therefore a factor in the rise to dominance of a few firms in industries such as aggregates, cement, bricks, paint, plaster and plasterboard and glass. Their large size allowed them to achieve substantial economies of scale. This, however, reduced the level of competition in parts of the materials sector which in turn led to a proliferation of product innovations rather than cost-reducing process innovations (Bowley 1960).

Bricks and nails were among the first and most standardized parts to be used in construction: they were produced in batch and later in volume production processes. The manufacture and standardization of basic materials were prerequisites to factory production of components. In order to achieve standardization, scientific examinations of building components were made, analysing them in modular categories, each representing a different attribute or function such as performance, structure, tolerance and installation characteristics. Prefabrication of sub-assemblies aimed to reduce costs, to increase speed of construction processes and to improve quality (White 1965). Prefabrication was therefore both a process and a product innovation. One benefit was that materials wastage on sites was reduced with the erection and assembly of prefabricated components when compared with handicraft production methods.

A distinction could be made between innovation in two types of prefabricated products: those produced for a specific building only after the design had been completed (production-to-order) and those which were produced without prior knowledge of the design or type of building in which they would be used (production-to-stock). The latter represented a further step in attempts to produce industrialized products for construction which could be purchased off-the-shelf and combined with other common products to form buildings (Kendall and Sewada 1987, pp. 7–19).

Systems-building approaches in the 1950s and 1960s involved more extensive use of prefabricated components. They also involved attempts to introduce quality control, new relations with manufacturers, the use of programming methods for construction sequencing, together with new methods of documentation. At the same time, standardization was given a new impetus through the design of buildings on a grid, or modular basis. The aim was to coordinate the size of factory-made components with the design of buildings. This became known as 'dimensional coordination'.

The forces promoting industrialized construction did not emanate solely from building designers and users wishing to emulate and transfer ideas from the production of automobiles and other manufactured goods. Manufacturers played important roles in promoting industrialized systems through the development and marketing of new products and components. Economic forces from within the construction process itself spurred the search for new methods of industrialized construction. For example, contractors realized that the prefabrication of standardized parts could cheapen components, reduce on-site labour requirements and speed up the construction process, and at the same time potentially provide buyers with higher quality products, because factory tolerances were tighter than those achievable on-site.

During the 1950s, the use of industrialized construction processes developed further through the introduction and widespread diffusion of new construction equipment—particularly lifting and materials handling machinery. Equipment for handling bulk materials, such as earth-moving plant, had the greatest impact on civil engineering. However, during this period, the tower crane was the most important piece of equipment to affect building construction. Use of tower cranes led to the reorganization of construction work sequences to enable maximum utilization of plant and returns on fixed capital investments.

2.3. Engineering and production methods from other industries

The industrialized production of construction materials and components, together with new construction techniques, were stimulated by the adoption of engineering and production philosophies from other rapidly expanding industries. Crystal Palace illustrates how ideas from other industries had been used to find solutions to technical, aesthetic, organizational and labour problems in the construction of buildings and structures. Outsiders, who had little to do with the traditions of construction in their normal course of work, played an important part as innovators. For example, reinforced concrete was invented by another gardener, Joseph Monier, who patented a reinforced concrete beam after experiments to make concrete tubs for plants at Versailles (Bowley 1966, p. 17). Alexander Graham Bell became preoccupied with the invention of tetrahedral structures which were later used in 'space-frames'—three-dimensional structural elements constructed from lattice bracing systems. These provide great strength over long spans at a fraction of the weight of solid beams. Bell's innovative approach was influenced by his interest in networks and systems and his efforts to replace rigid structural members with wire cables—related, perhaps, to his excitement over the use of wires in the telephone system (Wachsmann 1961, p. 34).

New materials such as steel began to replace cast iron and, together with the invention of elevators and water pumps, facilitated the construction of multi-storey buildings. The Bessemer Process was invented in 1856, with the first steel sections suitable for building purposes rolled in 1884 (Russell 1981, p. 59). This coincided with the dynamic growth of Chicago in the 1880s and 1890s—the birthplace of 10- to 16-storey buildings called 'skyscrapers', the architecture of Frank Lloyd Wright and the Chicago School (Giedion 1967). The walls of skyscrapers were constructed from rows of identical, standardized, volume-produced windows, known as the 'Chicago window'. But skyscrapers would not have been feasible without the use of other recent innovations of the time, notably, telephones and elevators. The telephone made it possible to communicate between different rooms and apartments, and entry phones facilitated communications with the world outside on the ground floor (de Sola Pool 1977). Elevators provided rapid and effortless vertical transport between floors.

The engineering content grew with the introduction of new technologies linked to large span construction used in road and railway bridges, stations, exhibition halls (such as Crystal Palace), market halls, department stores, docks and factories. Giedion describes many such examples (Giedion 1967). The Brooklyn Bridge in New York (1867–1875) was one of the first structures to include the use of cables as structural elements, while the Firth of Forth Bridge in Edinburgh (1883–1889) and the Eiffel Tower in Paris (1887–1889) were early examples of space-frames with lattice structures. Space-frames were later used to cover the large spans required in airship and aircraft hangars. The use of lattice girders was heavily influenced by airship design, and they were constructed from volume-produced components. The erection of these and similar structures was simplified by the development of structural

Fig. 2.4. Space-frame connectors

connectors which could be used irrespective of the material from which the structural elements were made (Wachsmann 1961, pp. 22, 24, 29–44, 88) (Fig. 2.4).

Leading architects and designers were vociferous in advocating the modernization of construction through the adoption of practices used in manufacturing. They were directly opposed to the Arts and Crafts Movement, which sought to preserve traditional craft products and practices. The Modern Movement, including Constructivists, the Bauhaus and Gropius, le Corbusier, Mies van de Rohe and CIAM (Congres Internationaux d'Architecture Moderne) wanted to do away with what they called 'medievalism', the concept of handicrafts and traditional craft practices, and replace these with machine-engineered solutions. The Bauhaus set about trying to impose rational order defined by technological efficiency and machine production. It was argued that '*the machine is the modern medium of design*'. Craft was redefined as the '*skill to mass-produce goods of an aesthetically pleasing nature with machine efficiency*' (Harvey 1989, p. 24). Buckminster Fuller believed fervently in the idea of the machine and mechanization. He argued that the production of buildings could be carried out in similar ways to that of cars and other volume-produced goods. In criticizing craft production he argued that (Russell 1981, p. 178):

> *craft building—in which each house is a pilot model for a design which never has any runs—is an art which belongs to the middle ages. The decisions in craft-built undertakings are for the most part emotional—and are based upon methodical ignorance.*

The ideas of mass-producing buildings, or at least building elements, had its detractors. In 1928, the American Institute of

Architects (AIA) passed a resolution that it was '*inherently opposed to any peas-in-a-pod like reproducible design*' (Baldwin 1996, p. 21).

Despite opposition from various establishment institutions, the ideas of the Modern Movement spread rapidly across Europe and North America in the early part of the twentieth century. Le Corbusier argued that '*houses must go up all of a piece, made by machine tools in a factory, assembled as Ford assembles cars, on moving conveyor belts*'. The arguments for introducing the benefits of the machine age to construction were dramatically put by the Italian architect and Futurist, Sant' Ilia (Banham 1960, pp. 128–130) (Fig. 2.5).

Fig. 2.5. Sant' Ilia's futurist vision

The problem of Modern architecture is not a problem of rearranging its lines . . . but to raise the new-built structure on a sane plan, gleaning every benefit of science and technology . . . We must invent and build our modern city like an immense and tumultuous shipyard, active, mobile and everywhere dynamic, and the modern house like a gigantic machine. Lifts must no longer hide away like solitary worms in the stairwells, but the stairs—now useless—must be abolished, and the lifts must swarm up facades like serpents of glass and iron . . . just as the ancients drew their inspiration in art from the elements of the natural world, so we must find our inspiration in the new mechanical world we have created.

In America, in 1915, Wright designed a ready-cut, prefabricated timber frame system for building flats. This was known as the American System. Steel and concrete framed house designs were also produced. Le Corbusier's Dom-ino House was perhaps the most influential with its simple, standardized, slender frame, slab floors, flexible floor layout independent of structure, lightweight movable internal walls, and external non-load-bearing cladding (Fig. 2.6). Traditional Japanese housing design based on standard units of area (tsubos) defined by the size of two tatami floor mats strongly influenced the design. However, in the West, this was to result in completely new methods of construction and was to strongly affect design philosophy behind the systems-built schools of the 1960s (Russell 1981, pp. 125–159).

The presence and role played by clients, or project sponsors, was one of the crucial differences between systems-building and other methods of industrialized construction (Finnimore 1989, p. 6). In Britain, mass housing programmes of the 1960s

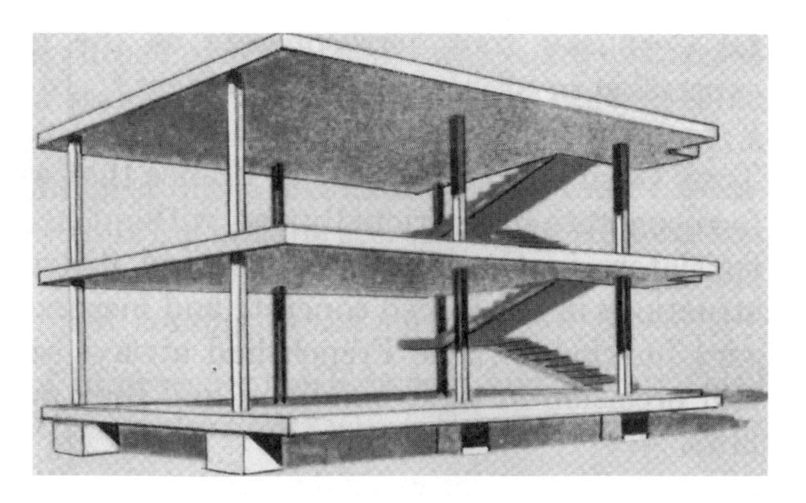

Fig. 2.6. Le Corbusier's Dom-ino House

facilitated the adoption of bulk purchase agreements for components which gave local authorities, consortia and large building contractors greater control over the supply of parts. It also provided a market for mass-produced construction materials and components. The use of standardized, volume-produced construction components in functional architectural designs reflected changes in the use of buildings.

2.4. Demand for new types of buildings and structures

Demand for new types of buildings and structures was stimulated by rapidly changing patterns and types of work together with processes of urbanization. By the early twentieth century, vertical high-rise cities sprang up where transportation lines intersected. Demand for tall buildings was in part stimulated by saving space in dense urban areas. Furthermore, skyscrapers permitted land owners to increase the rental available on a given plot.

In the USA, demands for new types of buildings were closely linked to the use of modern technologies such as the telephone. For example, the Pennsylvania Railroad was the first railway company to install telephones in place of telegraphs and, by 1910, ran a private telephone system comprising 175 exchanges with 400 operators and 13 000 phones linked by 20 000 miles of wire (Davies 1993, pp. 97–101). This had major implications for innovation in buildings. The development of private branch exchanges (PBXs) was not only an important step in the provision of services for large business customers who could now link different divisions at various locations, but PBXs also needed to be housed in special rooms within buildings. This is an early example of how more recent and dramatic developments in the use of public and private networks has led to the construction of 'intelligent buildings', discussed in Chapters 4 and 5.

In the early to mid twentieth century, managers in many different industries were organizing their space requirements to meet the needs of operating under the efficiency of scientific management and to imitate the Fordist production systems. The introduction of Ford's production line required reorganization on the shopfloor to accommodate dedicated machinery, the flow of components and stocks of parts needed for volume production. New, cheap factory buildings were required to enclose large spaces. Industrially produced large span sheds appeared to be the appropriate answer. After World War Two the need for large-scale planning, high-speed transport and high-density development became even more paramount, and in many respects the modernists believed they had found the solutions: their ideas were widely put into practice.

Offices and office furniture underwent considerable change between the 1880s, with the introduction of the typewriter, and

Fig. 2.7. Typical Fordist office layout

the 1920s, by which time the typing pool had become commonplace. Desks were redesigned and arranged in large open plan rooms so that supervisors could watch out for laggards (Fig. 2.7). Conveyor belts were even used to move paper into mechanical filing systems, thus removing the need for clerks to leave their desks (Forty 1986). Hospitals were planned so that nurses need only walk minimum distances—standardization and uniformity of hospital design was thought to contribute to the most efficient treatment of patients (Dickinson 1914). Similarly, schools and universities were planned to provide the most efficient 'throughput' of pupils and students. They were likened to factories, and designers were commissioned to think of ways to improve the 'yield' using new seating plans and room layouts, removing disorder and providing clean, centrally controlled buildings using standard designs. This was also intended to achieve more economical use of buildings (Cooke 1910, Callahan 1962). The ideas of labour saving and technical efficiency were extended to the home. Kitchens were designed to simplify work and standardized units introduced. Bathrooms were designed to provide hygienic, functional environments and were built from standard volume-produced components.

2.5. The role of the state in promoting innovation

Clients, planners and architects in local and national government, responding to new social pressures, actively promoted modernization in construction. The politicians, architects and administrators behind the drive to industrialize construction had little understanding about how the traditional industry worked, or why it was in many ways successful (Russell 1981, p. 202). Their attempts to replace traditional practices with modern methods often arose from pressures to overcome shortages of bricklayers and other craftsmen. In spite of such pressures, parts of construction, particularly house building in Britain, remain a predominantly brick based industry – evidence of the state's failure to push the sector onto a new footing.

In Britain, the social and economic policies of the Welfare State represented increasing state intervention in working class living conditions and a public interest in the relationship between technology, social and economic progress. This culminated in Wilson's policies associated with the 'white heat of technological progress' in the 1960s (Finnimore 1989). Parker Morris standards were, for a short time, widely adopted. This meant that public sector housing had to be designed to meet minimum space and quality requirements. The Parker Morris Committee published '*Homes for Today and Tomorrow*' in 1961. This was the first major review of social housing standards since the Second World War and it concluded that the quality of social housing was not keeping pace with living standards.

In twentieth century Britain, three great waves of publicly funded construction activity stimulated the use of new construction technologies: 'Homes fit for Heroes' after World War One; reconstruction after World War Two; and slum clearance and school building programmes of the 1950s and 1960s. In 1917, government committees investigating the production of housing found that bricklaying accounted for 31 per cent and carpentry for 26 per cent of total costs. Carpentry and joinery lent themselves to industrial techniques through mechanization, whilst bricklaying did not. The search for substitute techniques to replace bricks as a structural material had begun. New materials of steel, concrete and glass superseded more traditional methods of construction and a wide range of experiments in prefabricated housing was carried out. By 1948, 30 per cent of permanent local authority house completions were constructed using systems-building techniques (Finnimore 1989). The scarcity of labour and materials and the need to turn redundant war time armaments factories to new uses also influenced the choice of industrialized housing construction.

The British government became actively involved in the sponsorship of construction R&D, forming the Building Research Station (BRS) in 1921. The BRS was the first organization of its kind in the world. During the 1950s and 1960s, R&D in construction was strongly supported by both British government and industry. New research institutions were formed, and heavy emphasis was placed on research on construction processes. The Building Research Establishment (BRE) was formed in 1972 from the amalgamation of four separate laboratories: the BRS; the Princes Risborough Laboratory (established in 1926 as the Forest Products Research Laboratory, where research included the performance of lightweight components and jointing systems); the Fire Research

Station; and the Civil Engineering Laboratory. The aims of the BRE were to carry out research into building and construction and the prevention and control of fires, and to serve the needs of central government in its responsibilities for maintaining and improving the performance, safety and economy of buildings.

In the early 1960s, British governments had set increasingly higher annual output targets for housing construction and there was concern whether these could be attained in the face of perceived shortages of site labour. Many continental systems were examined with a view for use in Britain. Systems-building practices were adopted with the aim of reducing requirements for on-site craft skills by increasing the use of factory-produced components and on-site mechanization. The view was widely held that systems-building would rapidly become the dominant mode of house building, especially in the public sector. Industrialized techniques were used extensively in the period 1962 to 1967, especially for the construction of high-rise apartment blocks (Merrett 1979, p. 89). During that period, the production of flats exceeded production of houses. In 1965, the government recommended the use of new materials and construction techniques, including dry processes, increased mechanization of on-site processes and manufacture of large components under factory conditions. Improved management techniques, closer links between design and production, and better control over the selection and delivery of materials were also recommended (Finnimore 1989).

The end of 1960s style systems-building and the belief in industrializing construction through the volume-production assembly-line approach was accentuated in Britain by the failure of Ronan Point, which partially collapsed after a gas explosion on 16 May 1968. Ronan Point was one of 26 blocks of flats in Greater London built using the Larsen–Nielsen system. This system had been developed in Denmark and used extensively there, although rarely for buildings of more than six storeys; it had been banned for structures above six storeys in the USA. Evidence given at the inquiry after the explosion concluded that the joints holding some of the panels were not sufficiently strong and that there were deficiencies in workmanship and supervision during its construction. Furthermore, evidence emerged that system-built housing was probably neither cheaper nor quicker than traditional construction. The prefabrication of large structural elements used in 1960s high-rise housing frequently resulted in leaking joints and problems of condensation (Russell 1981, pp. 446–455).

In addition to extensive technical failures, there were growing objections to high-rise flats on social grounds. In Britain, the pressure for housing had led to far more families living in such blocks than had been envisaged. The strain imposed on parents and the developmental problems for children associated with living in wind swept, aggressive and oppressive environments were becoming increasingly recognized. As a result, system-built high-rise flats became discredited in Britain and this rejection was reflected in 1967 by the ending of additional subsidies per flat in respect of blocks over six storeys high. After 1966, there was a rapid fall in the number of tenders approved for high-rise flats (Merrett 1979, pp. 130–131).

2.6. How successful was construction in the machine age?

Attempts to introduce industrialized construction techniques originated from the manufacture of simple construction materials and components, and were later inspired by the success of volume-production methods such as those used in the automobile industry. Manufacturing processes involved the concentration of materials, fixed capital and labour in one or more places. They successfully demonstrated improvements in efficiency over scattered craft production found in many traditional industries, including construction. Manufacturing provided three main advantages over craft:

(a) economies of scale, when the cost per unit produced dropped more quickly than production costs rose as the volume of materials processed increased;
(b) technical possibilities to develop and deploy capital equipment;
(c) the opportunity for tighter managerial control.

These advantages were exploited by Henry Ford whose adoption of scientific management and invention of the mass-assembly line facilitated the production of high volumes of standardized products made from interchangeable parts with simple fixings. The same gauging system was used through the entire manufacturing process, driven by savings on assembly costs. It was these innovations that made the assembly line successful. It employed unskilled or semi-skilled workers on expensive, dedicated machinery. Design and management was carried out by narrowly skilled professionals. Moreover, because the machinery was so expensive, firms could not afford to allow the production line to grind to a halt. Buffers, such as extra supplies of materials and labour, were added to the system to assure smooth production. Furthermore, because changing machinery to produce new products was even more expensive, producers kept standard designs in production for as long as possible. This resulted in consumers benefiting from lower costs but at the expense of variety (Womack *et al.* 1990, p. 13).

The stated aim of industrialized construction was to realize similar benefits to those enjoyed in the production of manufactured goods—raising efficiency by rationalizing the process through the application of scientific methods. There has been no systematic measurement of the overall gains resulting from the use of prefabricated components, but evidence suggests that systems-building did not raise overall productivity and was rarely cheaper or much quicker than traditional construction techniques. Nevertheless, the British school building programme in the early 1960s is often heralded as a success for solving some acute accommodation problems within tight budgetary constraints (Harvey 1989, p. 114). Studies of the effectiveness of systems-building have shown that whilst labour time was saved in the erection of structures and external cladding, a considerable part of the labour-intensive finishing of internal work was left to be performed in traditional ways. Construction continued to suffer the problems of craft production: high production costs which did not drop with volume. The craft workers who were essential for finishing buildings and rectifying faults were usually employed by small local firms. While these workers may have been aware of technical problems caused by bad design or poor workmanship, the management and organization of the process did not provide space for them to feed this knowledge back to those with responsibility for such matters. Moreover, these firms were far too small to engage in the development of their own technologies or to train new generations of workers with appropriate skills.

The invention of standardized, interchangeable prefabricated construction components had some similar effects on building work to those experienced in automobile production and manufacturing. Just as the Fordist production system swept aside craft car producers (except for those employed in the re-work shops), so industrialized construction techniques eroded traditional craft skills. Ending craft practices was one of the goals of industrialized construction. Tasks were divided and sub-divided, craft control was replaced by new management practices and the pace of work was often dictated by the need to maximize the use of equipment such as tower cranes. Employment for construction workers became casualized in a similar way to that of car production line workers who were treated as interchangeable parts and taken on or laid off as and when they were needed. Immigrant workers were used to supplement indigenous labour, particularly in the most menial tasks, in the same way that 'guest workers' have been used on volume-production lines in manufacturing industries.

Large construction projects were oriented towards assembly and away from the use of handicraft skills. By the 1960s, the British government was recommending that teams be trained to work on long repetitive runs. But while a straightforward de-skilling process was evident in parts of construction where traditional craft skills were being divided, new skills were also required in work with new materials and machinery. Furthermore, many parts of the sector were relatively untouched by industrialized methods. Small-scale building firms continued to produce housing using traditional methods, sometimes building only one or two units a year. Repair and maintenance and finishing work relied heavily upon the expertise of craft operatives who were sometimes called upon to bridge the gap between buildings constructed using old technologies but which included new installations.

The problems of emulating mass-production techniques were partly associated with coordinating the huge number of parts which had to be assembled in complex parallel processes. Depending upon how parts are counted, a car is assembled from around 10 000 components whilst a house may be constructed from as many as 40 000 components. The physical nature of the construction process as a constraint creeps back into the argument. Construction never moved far from the use of prefabricated components on demand for specific projects, to prefabricated components off-the-shelf for any project. Furthermore, the component systems were often 'closed' and inappropriate for interconnection with systems produced by other manufacturers. Each construction site was treated as a new factory employing mobile gangs of workers. Fordist mechanisms of control were difficult to adopt, partly because policing the ever-changing work area was an onerous task in itself. Nevertheless, work-study, piece-work and bonus payment methods were used widely on large projects.

Finally, the markets for mass-produced buildings were often not perceived as being stable enough to warrant the huge investment costs which would be required to tool up the industry to produce components in factories. In manufacturing, firms had secured large continuous markets, which they were able to organize and control with some degree of success. The housing market is often compared to that of cars or other consumer goods, but while the housing market is huge, it cannot be organized easily. This is partly because of the long-lived nature of the product, forms of land ownership, and because of a variety of methods of planning, financing and consumption. Furthermore, the functional design of many prefabricated components may not always satisfy consumer desires. Lessons from 1960s high-rise dwelling construction show that the design and construction of buildings needs to result in products which are socially acceptable—this proved to be difficult to achieve using standardized volume-production techniques.

2.7. Summary

The construction sector followed two paths of development from the mid nineteenth century to the early 1970s: craft production and industrialization, typified by attempts to use volume-production methods. The main features of each are shown in Table 2.1.

Two different types of innovation were evident in the machine age: those which changed the product and those which affected processes in terms of costs and availability of inputs (Bowley 1960). The most important of these were the new buildings and

Table 2.1. Two paths of development, 1850s to 1960s

	Craft	**Industrialized**
Process	Handicraft	Development of *in situ* assembly
Markets	Small-scale traditional markets: residential and repair and maintenance	Large-scale projects, new markets: construction of infrastructures, new mass-housing, slum clearance, factories, offices, schools and hospitals
Product	Bespoke, made from basic materials	Standardized, made from factory-produced components
Type of firm	Small, using locally available resources	Large, national or international; suppliers sometimes monopolizing regions or part of oligopoly operating cartels using chains of smaller subcontractors
Skills	Craft trades demarcated by skills associated with the use of particular materials	Specialized, narrow technical skills, fragmentation of old craft skills, growth of new skills associated with new materials and techniques
Learning	Cumulative	Application of scientific and engineering knowledge
Innovation	Unstructured, informal	Structured, formal R&D
Technological change	Incremental changes, adaptation of 'tried and tested' techniques based on the use of traditional materials	Major changes such as prefabrication and the development and use of new materials, construction plant and equipment
Organizational change	Minor adaptations to traditional craft forms	Adoption and adaptation of methods used in manufacturing sectors

structures that accommodated new functions for clients and users.

At the same time, traditional craft forms of production and the structure of the building industry itself constrained the diffusion and further development of innovative approaches. The organization of construction processes was inappropriate for meeting the requirements expected of a modern industry. The organization of production was a major constraint to innovation and there was a remarkable lack of cooperation between the various parties involved in the design and erection of buildings. There was also an absence of informed rational decision making by building owners. The development of separate and specialized design professions helped to bring about divergent interests between engineers and architects and the organization of design had no built-in mechanism to bring about innovation in the absence of external stimuli from clients or materials' producers.

Furthermore, architects, with overall responsibility for the designs, engineers who designed structures and builders who erected them, were all isolated from one another. Innovation by contractors was impeded by the small size and fragmented nature of local markets. There was an absence of competitive pressures on non-innovating firms owing to the protection afforded by transport costs and the lack of internal dynamics which necessitated an expansion in markets to increase the willingness to innovate. The scope for market-expanding innovation by contractors was limited by their inability to specify their products while their scope for cost-reducing innovation was limited by the variability of the type of work they could obtain. The form of contract, the cost of carrying out experiments, lack of information on costs relevant to the development of cost-reducing innovations and obstacles created by bylaws and codes of practice which enshrined procedures based on past experience all hindered progress (Bowley 1960, Bowley 1966).

For these reasons, the craft and industrialized paths of development continued along their respective trajectories for much of the twentieth century. During the early and mid twentieth century, large contracting firms and building materials' producers and suppliers emerged and began to monopolize parts of the sector. Standard forms of contract were devised to manage the relationships between the many participants. Professional and managerial tasks became narrowly specialized. Design became further divorced from production. Industrialized techniques were 'locked-in' on standardized prefabricated component systems, the use of machinery for lifting and transportation, and the adoption of management techniques first used in manufacturing. By the late 1960s and 1970s, however, construction was again at a crossroads. Industrialized systems had failed to deliver the range of choice users wanted and craft processes were too time-consuming, expensive and could not produce the new engineered buildings and structures required. This time the choice was not simply between craft or industrialized techniques.

3. Machines for living

Buildings provide shelter from the external environment and facilities to support daily life at home, in the work place and for education, health, travel and leisure. Two elements are needed to provide these facilities: passive components, derived from the architecture and engineering of structures, envelopes and fabric of buildings; and active elements including machinery and electrical systems which are fitted inside them. These provide the means for conditioning and controlling internal environments and an internal infrastructure for moving people, goods and services within the built environment. Other constructed products such as transport and utilities infrastructures can also be defined in terms of their passive or structural elements and active, servicing systems.

This chapter addresses technological innovation in active mechanical and electrical systems, arguing that they have become increasingly complex with the growth in demand for buildings to support new facilities and better control over internal environments. A separate industry has emerged to design, engineer and produce mechanical and electrical systems for buildings. This has a number of distinct features when compared with architectural and structural engineering. The chapter begins with an historical review of innovation in mechanical and electrical systems, followed by an analysis of the introduction of vertical transportation systems within buildings. Finally it considers innovation in control systems needed to manage increasingly complex technologies installed within buildings and the internal environments such equipment is supposed to condition.

Fig. 3.1. The building as machine – The Arch, Paris (mid 1980s)

The development and diffusion of mechanical and electrical systems in buildings and structures represents one of the major areas of innovation in the built environment during the machine age. Over the past two centuries buildings and structures have become more like machines themselves. For example, the Eiffel Tower was produced by steam cranes that climbed the tracks which also formed part of the structure, and the tower's feet were embedded in hydraulic presses that could be activated to adjust the level of each leg as the tower grew. As more machinery was installed in train stations, hospitals, banks and elaborate opera houses, they became known as *facilities*, rather than buildings in the traditional architectural sense (Fig. 3.1) (Peters 1996, p. 352).

Architectural design and engineering of what were essentially passive elements of buildings aimed at providing comfort can be traced back to the empires of Greece and Rome, where the basic human need for warmth was met through the use of simple central heating systems. Yet in Britain and other parts of Europe the centuries between 450 and 1500 were marked by a step back from the more innovative Roman designs like the hypocausts. During this time, buildings were heated by fires burning in hearths. At first there were no outlets for smoke, but later, holes in roofs were provided. The central hearth and hole in roof method of heating was replaced by the wall fireplace and chimney system; the earliest known examples of which were found in European castles in the eleventh and twelfth centuries. Fireplaces and chimneys were improved and their use became widespread between the fifteenth and seventeenth centuries.

General knowledge about rudimentary requirements for light, air, warmth, clean water supplies and sanitation developed through the Middle Ages and systematic studies of basic needs began during the scientific revolution. The principles behind modern approaches to the development of active mechanical and electrical systems in buildings started with the scientific study of human physiology and metabolism in the eighteenth century. The invention of the thermometer by Fahrenheit in 1724 was of critical importance, facilitating the accurate measurement of temperature, a precursor to the application of scientific knowledge and development of engineering expertise for temperature control. This was followed by the evolution of thermodynamics and fluid mechanics in the nineteenth century.

By the early nineteenth century, measurement, specifications and designs to meet basic comfort, safety and functional needs were carried out on a scientific basis. Later, the combined effects of temperature and humidity were analysed. The study and practice of environmental engineering grew into a distinct discipline and by the end of the century specialist professional institutions were formed, such as the American Society of Heating and Ventilating Engineers and the Institute of Heating and Ventilating Engineers in Britain. More sophisticated measurements of comfort were devised through experiments carried out in the USA in the early twentieth century. For example, in 1900, C.F. Marvin of the US Weather Bureau devised psychrometric tables for obtaining the vapour pressure, relative humidity, temperature and dew point of air. The science, technology and engineering involved in controlling internal environments became more complex as attempts were made to calculate the combined effects of variations in acoustics, lighting, air quality, humidity and ventilation together with temperature. Results from these experiments were compiled in standard 'comfort charts' which could be used by designers specifying equipment for different types of buildings in various locations (Billington and Roberts 1982).

Furthermore, the Industrial Revolution resulted in new markets and an expansion in the scope of mechanical engineering from satisfying basic comfort needs to the provision of facilities for production equipment, such as steam, compressed air and, later on, electric power. Similar systems were installed to facilitate leisure activities. By the middle of the nineteenth century the provision of environmental control and infrastructures to support work and leisure activities within buildings had become inextricably linked together.

3.1. Innovation in mechanical and electrical systems

From the mid nineteenth century onwards, innovation in mechanical and electrical systems occurred in several separate branches of technology. Most were associated with the introduction of new types of mechanical equipment, together with electric power. In terms of comfort control, engineering principles were first applied to heating systems, followed by refrigeration and ventilation. Blocks of ice and freezing mixtures were in widespread use for food preservation from the mid nineteenth century. Gas liquefaction refrigeration (vapour compression) was achieved in the late nineteenth century, together with the first commercial production of household refrigerators. Electrolux pioneered the use of freons in continuous absorption apparatus used for refrigeration equipment between 1900 and 1910. This drew upon the expertise of chemical as well as mechanical engineers.

Electrical systems began to be used from the beginning of the twentieth century. New factories and offices required power, lighting, heating and ventilating, while food transport and storage required refrigeration facilities.

3.1.1. Heating and ventilating

Plumbing was the first trade to emerge. Plumbers, working with lead, constructed active mechanical systems dating back to the Middle Ages. Their work was associated with sanitation, hot and cold water supply, and lead work on cathedral roofs and large stately homes. Plumbers were involved in much of the work to provide water and steam to factories during the Industrial Revolution. Boiler making grew as a separate activity as industry expanded. Table 3.1 illustrates some of the main innovations in heating. The demand for many of these was associated with controlling and conditioning the environments in new buildings constructed to accommodate the expansion of industrial and commercial activities.

Expertise within the plumbing trade shifted from a craft orientation towards an engineering approach and as it did so plumbers began to extend their field of activity into new areas, such as the provision of ventilation. They also worked with a

Table 3.1. Innovation in heating

Date	Innovation
Early C18th	Piped hot water heating first used
Early C19th	Warm air heating used commonly in Germany, Austria and Russia. Franklin's iron stove isolated smoke and gases from breathable air, thus resolving major conflicts in the need for warmth and ventilation
1827	Friction match invented (Billington and Roberts, 1982, p. 73)
1852	Watson and Slater patented a method for using electricity for heating (commercially used at end of C19th)
Late C19th	Gas and oil fired heating replacing coal
Late C19th	Steam heating in common use
1906	Introduction of Nichrome wire permitted electrical elements to operate at red heat
1925	Large-scale thermal storage first used in Britain
1930s	Solenoid or magnetic radiator valve operated by room thermostat developed in the USA and used in Britain
1930s	Mechanical clock control of boilers became common in Britain
1930s	Electric power and control of gas fired boilers invented in Britain
1950s	Thermostatic control of small domestic boilers was available in Britain
Mid 1950s	Time switch devices became widely used on boilers
1960s	Use of thermostatic valves became widespread across northern Europe
1960s	Central heating became common in British housing

wider range of materials including copper, cast iron, alloys and, by the 1950s, plastics. Table 3.2 illustrates some of the major innovations in ventilation, which were initially seen as an extension of plumbing work. However, as demand for ventilation, refrigeration and, later, air conditioning grew, these became specialist technical engineering trades in their own right — modern building engineering service industries had emerged.

The growth of ventilation know-how was closely associated with problems of illuminating dark interiors using naked flames as well as removing smoke caused by heating and cooking. One solution relied upon passive architectural elements such as the provision of louvred lanterns with vertical windows that provided both natural lighting and ventilation. A similar concept is used in the design of modern atria which can be opened to

Table 3.2. Innovation in ventilation

Date	Innovation
C18th	Mechanical ventilation developed—steam-powered fans were used in mining
1860	Sturtevant Company began manufacturing centrifugal fans
Late C19th	Mechanical ventilation became common in large public buildings
Late C19th	Growing awareness of the need for adequate natural ventilation of buildings, particularly in dwellings (ventilation was needed where open fires were used)
Early C20th	Electrically powered fans first used
1900	Work study on efficiency in schools recommended that buildings should be adequately ventilated to reduce drowsiness during lessons (Cooke 1910)
1936	First electric window fan made by Vent-Axia
Late 1950s	Mechanical extractor fans used in flats in Britain
1960s	Pre-balanced extractors used for toilets in office blocks and flats

exhaust smoke in the event of fire. In hot climates, the opposite was necessary—the need to circulate cool air.

Whilst requirements to create comfortable and healthy environments were a major factor in the development of ventilation systems, two other conditions were also important. First, the need to maintain structural integrity, protecting buildings and structures from decay caused by damp internal conditions and, second, preservation of contents such as harvested crops or industrial materials (Fitchen 1986, p. 189). Ventilation became more important as industry developed, with the need to reduce respiratory problems and other health hazards caused by foul air, dust and noxious gases, particularly in mining. In spite of advances in the engineering of heating and ventilation systems by the end of the nineteenth century, the general public displayed little interest in, or knowledge of, the benefits of proper ventilation. These were widely published in popular magazines with the aim of reducing problems of inadequate or non-existent ventilation which were crippling city dwellers (Banham 1969, ch. 3) (Fig. 3.2). Britain lagged behind the USA in the use of new technology for heating and ventilating. British government offices in Whitehall were still being heated by open coal fires in the early 1950s and there was little concern over energy conservation in either country until the 1970s (Atkinson 1993, p. 133). Issues associated with public awareness of appropriate ventilation continue today, 100 years

Fig. 3.2. Advertisements for technologies to protect public health, 1900

after the first major public campaigns. A range of problems relating to respiratory difficulties in housing and work places include the need to protect against deadly carbon monoxide gases and to reduce the problems of 'sick building syndrome'. In the 1990s in Britain, asthmatic conditions caused by fungal spores in damp, humid, poorly ventilated housing continue to cause problems.

The increase in the installation of mechanical systems in buildings led to the emergence of new design and engineering activities which differed from more traditional building trades such as masonry, bricklaying and carpentry in their bias towards technical engineering skill. From the outset, designing, making and installing mechanical systems in buildings had closer links with fabrication workshops and factories. Many elements of their work were therefore more industrialized than those of traditional building trades responsible for the construction of the passive elements of building structures, facades and internal fabric.

3.1.2. Electric power and light

There was a great period of inventiveness and experimentation with electricity in the late nineteenth and early twentieth centuries (for example, see Josephson 1961, Dunsheath 1962, Haustein 1980, Hughes 1985, Forty 1986). Electricity superseded other forms of energy for the provision of light, power to production plant as well as to mechanical building equipment such as fans and, in many cases, heating. Electric power replaced steam and water power in factories, leading to major changes in the design and construction of buildings and in the type of mechanical and electrical systems installed within them (Fig. 3.3) (Devine 1983). Between 1880 and 1930, the production and distribution of mechanical power in factories rapidly evolved from water and steam, which drove shafts and belts linking many machines, to electric motors which drove individual machines. This signalled a shift from the direct use of raw energy in the form of coal and water power to the use of processed energy in the form of internal combustion fuel and electricity.

The major advantages of electric power in factories were a reduction in energy required to drive machinery, resulting in greater output per unit of capital and labour input, and increased control over individual machines. In 1900, steam power provided around 80 per cent of mechanical drive capacity, but by 1929 the use of electric motors had become so widespread that they accounted for about 78 per cent of total capacity for driving machines (Devine 1983). The substitution of electric power for steam was associated with major improvements in factory organization, resulting in large increases in productivity. Among the first to use electric power were the clothing, textiles and printing industries where a clean environment, steady power,

Fig. 3.3. From steam to electric power in factories

speed and ease of control were important. In the early stages of implementation, electric motors facilitated improvements in the quality of work but the method of providing power to machinery on factory floors was left unchanged. Electric line shaft drive systems used the same countershafts, belts, pulleys and clutches, which drove machines in the same way as the traditional direct drive steam or water powered shafts. This system was vulnerable to breakdowns that affected the whole factory, but it was used because engineers were familiar with the problems associated with shaft and belt distribution. The power distribution system was therefore initially left unchanged; the new technology of electric motors was juxtaposed upon the framework of the old technology of steam power. However, electric group drive and individual drive units were soon introduced. These reduced the risk of total factory stoppages as well as the level of airborne dirt associated with drive shafts and belts. Further improvements in efficiency resulted because machines could be operated independently.

The change in distribution from direct drive to distributed electric power had a major impact on the structure, layout and servicing of buildings. Electric power sources could be located separately from factories, removing the constraints of direct drive on the size and configuration of individual buildings. Factory buildings became lighter and cheaper to construct because heavy line shafts, countershafts and pulleys did not need to be supported in roof spaces. Moreover, the absence of overhead mechanical power transmission allowed improvements in lighting, ventilation and cleanliness. Machinery no longer needed to be grouped and placed relative to shafts—it could now be arranged to maximize throughput from a particular manufacturing sequence. Individual machines could also be moved with little interference to the operation of other machines.

Devine's account (Devine 1983) of the development of electric power illustrates two general points concerning technological change in the construction and use of the built environment. First, new technology was initially juxtaposed on the framework of an old one: there are many examples of this in the application of digital electronic systems discussed later. Second, the full implementation of electric power and distribution resulted in an increase in flexibility in the utilization of space, facilitating the design of new types of factory buildings. Similar results have occurred more recently with the introduction of digital systems in offices and factories.

Analysis of the invention and incorporation of electric light indicates a number of further points concerning innovation in technical systems, which have general applicability and help to explain the nature of more recent changes in mechanical, electrical and electronic systems. Prior to electric light, increased lighting power was derived from open flames: candles and torches preceding piped gas. The advent of electricity and the incandescent lamp revolutionized lighting and made it possible to use space in new and different ways. It brought with it a new age of public electricity supply. Among the advantages of electric light were that it produced little heat compared with naked flames, and no soot. It needed no trimming or servicing compared with gas lamps and it could be installed in many restricted spaces. Table 3.3 lists some of the major innovations in lighting.

Edison formed a company to produce electric light bulbs in 1878. By October 1879 he had lit the first lamp, and by December of that year was demonstrating the first incandescent lighting system. The first public mains was opened in 1882 in

Table 3.3. Innovation in lighting

Date	Invention/Innovation
Up to mid C19th	Combustion — candle, oil, gas, etc.
1877	Jablockov invented carbon arc lamp
1879	Edison invented carbon filament lamp
1881	Temperature radiation bulb
1900	General Electric developed the electric meter, a major contributor to improved efficiency in use of electricity
1905	Metallic filament lamp invented
1909	Gas lighting declined in use as it was gradually replaced with electric lighting
1911	Tungsten filament lamp invented
1915	Gas-filled coiled filament lamp invented
1920s	Electricity promoted by the electrical development association as the 'healthy, labour-saving fuel of the future'. Its use for heating, cooking and cleaning, etc. became widespread
1921	Luminous condenser effect discovered
1930	Gas discharge lamp invented
1932	Mercury lamp invented
1935	Wound coiled filament lamp invented
1938	Krypton lamp invented
1939	Fluorescent lamp invented — first mains voltage tubes marketed in 1940
1950	Major improvements to wound coiled filament lamps
1959	Halogen lamp invented

the USA. Edison's prime objective was to develop an incandescent light which was economically competitive with gas and which could deliver cheap illumination to millions of households. The purpose was therefore cost-reducing and market-expanding technical change. It meant breaking with series wiring configurations used at that time in arc lights, and developing a parallel wiring system. The lamp was the leading edge of Edison's systems approach. Having perceived the function of this component in the system, the design of generators and other equipment was defined by the characteristics of the lamp. Components were then developed in parallel. Edison's approach was to ignore disciplinary boundaries: he was problem- not discipline-oriented.

The development of electric light and power required close cooperation between the inventor Edison, the systems manager, Insull and the financier, Mitchell. All shared a single common vision and each was an entrepreneur with strong organizational flair and the ability to conceptualize systems. They determined the price of lamps, initially making a loss in order to expand the market (Houstein, 1980 and Hughes, 1985).

In the 1920s and 1930s, many innovations in lighting were developed by the film industry, including the use of flood, spot and colour lights. But these were too large and powerful for immediate commercial application in buildings. At the same time, manufacturers did not seem driven to innovate further and, in spite of their apparent lack of interest in other mechanical and electrical services, it was architects who pushed manufacturers into developing new forms of lighting for buildings. The illumination of internal spaces probably had an aesthetic appeal to architects whilst the engineering of environmental services and control systems did not. However, new and more powerful lighting brought increased heat loads requiring more air conditioning to cool buildings. The provision of mechanical and electrical services in buildings was growing in complexity and developments in this one area were having repercussions on others. Electricity had begun to make a big impact on the design, engineering and construction of more traditional, passive elements of building structures, facades and internal finishings—for example, the introduction of suspended ceilings for the installation of lighting and air conditioning.

By the mid twentieth century lighting of commercial and public buildings was usually distributed on a modular grid basis on a horizontal plane. The emphasis was on uniformity of light in each area. Lighting systems were designed from standardized mass-produced components, purchased off-the-shelf. The design, production and use of lighting was typical of the standardized, volume production and consumption patterns found in other industries (such as automobiles) at that time.

3.1.3. Electrification in housing

The advent of electric power also brought with it major technical changes in housing. Not only did it provide new, safer forms of light and heat, replacing gas and incandescent sources, it also provided a source of power for electric motors. From around 1915 onwards, electrification in the home saw the introduction of new electrically powered machines and gadgets for domestic use. The labour-saving house was a cliché of the 1920s onwards, with the introduction of the small electric motor. It was in part a response by the emerging middle classes to the perceived shortage of domestic servants (Forty 1986, p. 118). The main aim was to replace domestic servants with machines, so that one person could look after the whole house with time to spare for other activities (Hardyment 1988, pp. 177–179).

The electrification of buildings was accompanied by major upheavals both in physically locating systems within homes and offices, etc. and in terms of the ways in which they were integrated within new functions in daily lives. In Britain, the Electrical Development Association (EDA) played an important role in creating the conditions necessary for the expansion of markets for electrical systems and products (Fig. 3.4). Market penetration probably took around 25 to 30 years, from around 1900 onwards. Adrian Forty (Forty 1986, p. 189) argues that in the mid 1920s, the conditions were right for demand for electrical appliances to grow. These included a cheap supply of electricity, cheap and reliable appliances and the installation of a distribution and wiring system. In Britain between 1918 and 1939, the extension of mains electricity supply to both new and existing houses meant that the proportion of households connected to the mains rose from 6 per cent to about 66 per cent. Many houses were equipped only for lighting during the inter-war years. Those houses that had electricity for power often only had one five amp socket. Probably only one third of all homes had more than two electric sockets, and they were principally newly built houses in the upper price bracket (Forty 1986, p. 189). Moreover, people were fearful of the consequences of wiring up their homes. Anecdotal cases known to Adrian Forty include two ladies who anxiously kept plugs in all the electrical sockets to prevent electricity from leaking out; another elderly lady was said to be terrified by the installation of even an electric bell lest the workman be killed in the process.

The need for labour-saving, hygienic, safe, comfortable and secure homes has been a priority for many decades. Manufacturers, installers and housebuilders have successfully integrated many mechanical and electrical systems and products into the home which have improved standards of living and changed the expectations of occupants. These changes took time and were often closely related to changes in design and construction of new housing, particularly in bathrooms,

Fig. 3.4. Advertisements promoting the benefits of electricity in the home

kitchens, utility rooms and living rooms. Houses were designed for new styles of living with modern technologies. This in turn spawned new design services such as kitchen planning. A relationship evolved between the design of active systems and passive elements in housing.

3.1.4. Air-conditioning

Air conditioning is a means of providing internal comfort using mechanical equipment. It involves the control of air temperature, purity, humidity and movement — a certain amount of draught is more comfortable than still air. Such systems were first developed in the early twentieth century by Willis Carrier, a North American engineer who integrated the technologies of heating, cooling (refrigeration) and ventilation. Carrier had worked for Buffalo Forge Company where he was involved in designing heating plant, systems for drying lumber and coffee, and forced draft boilers. He realized that existing rule-of-thumb techniques produced over-designed and inefficient systems and set out to develop a more scientific approach combining the three areas of expertise (Ingels 1952). This fusion of technology provided a more cost-effective way of controlling internal environments. Air conditioning also allowed for space to be used in new ways, for example in the design of deeper buildings with cellular space. Major innovations in air conditioning systems are shown in Table 3.4.

Economies were achieved by reducing the number of pipes, conduits and control devices that would have previously been required in separate systems. In the USA, air conditioning became one of three elements to be combined together in what became known as suspended ceiling systems: the other two were acoustic tiles and fluorescent lighting. These became popular in skyscrapers and by the late 1940s, standard kits of parts were available. But there was a difference between the American and European approaches to environmental engineering which resulted in the USA gaining a lead in the integration of mechanical and electrical systems engineering for buildings. Banham, continuing his theme about the separation of architecture and mechanical and electrical systems engineering, argued that, in air conditioning, American engineers had developed a technology that would make the modern style of architecture 'habitable by civilised human beings' (Banham 1969, p. 162). Meanwhile, European modern architects were attempting to devise a style that would 'civilize technology' — a style in which such reliance upon mechanical control would be less necessary. Architects dominated the European approach whilst engineers began to play a more central role in the USA.

3.2. Vertical transportation systems

During the latter part of the nineteenth century the invention and development of elevators represented another example of rapidly changing equipment to be installed within buildings. The

Table 3.4. Innovation in air-conditioning

Date	Innovation
Late C18th	Industry required air-comfort systems
Late C19th	Air-conditioning developed for use in hot climates
Early C20th	Willis Havilland Carrier, leading inventor of air-conditioning systems in the USA, produced many innovations
1900–1930s	Expanding demand for air-conditioning: refrigeration and ventilation in ships, regulated hot air for drying coffee and tea, bulk cooling in breweries, control of dust in tobacco factories, control of mould growth on celluloid, fibre humidity control in weaving, ventilation in mines
1920s	Demand grew from the need to control climates in soundproof broadcasting studios
1920s–1930s	Production of conditioned air throughout large buildings was successfully achieved
1925	The Rivoli in New York was the first movie theatre to offer 'air-cooled' comfort, introducing the general public to an improved atmospheric environment
1930s	Air-conditioning systems recognized to be necessary in all large buildings, e.g. hotel dining rooms and ballrooms, theatres, and in Pullman cars
Mid 1940s	Packaged air-conditioning modules introduced
1950s	Diffusion of air-conditioning systems into smaller buildings and dwellings

idea of constructing taller buildings had stretched the imagination of entrepreneurs and developers for many years. Interest in the idea grew during the period of Great Exhibitions in London, New York and other major cities in the mid nineteenth century. Their realization was, however, constrained by the technical limitations of moving people and goods through great heights. Traditional stairways were clearly not feasible for moving up or down more than a few storeys.

The earliest types of elevators were usually operated by hydraulic power. Petroski includes the invention of elevators in his description of design and innovation in tall buildings. He argues that elevators were often prone to failure due to fractures in hydraulic cylinders or breaking lifting ropes (Petroski 1996, pp. 194–197, 200–204). Concerns over safety limited their use. James Borgardus, an American mechanical and structural engineer, proposed a 300 ft high cast iron tower for the New York Exhibition of 1853, in which visitors would be transported by a steam-powered elevator. Borgadus's design was rejected because of fears over safety. However, at the same exhibition,

Fig. 3.5. Otis' demonstration of the elevator safety device at the New York Exhibition in 1853 (Source: Petroski 1996, p. 195)

another mechanical engineer, Elisha Graves Otis, demonstrated a new safety device which checked the fall of an elevator if the supporting ropes were broken (Fig. 3.5). This invention reduced the risks of mechanical elevation systems and public fears over safety subsided. By 1857, a hydraulic passenger lift was installed in a five-storey store on Broadway and the idea of vertical transportation systems soon became commonplace. The technicalities of using hydraulic systems imposed height limitations: one of the highest was installed in the 21-storey Flatiron Building designed by Chicago architect David Burnham and completed in 1902 in New York. This was the tallest building in the world at the time and one of the first built with a steel frame structure—it heralded the era of skyscrapers.

In the early part of the twentieth century buildings rose much higher than the rather squat Chicago skyscrapers and these required new types of elevators. The Woolworth Building (1913) had two of the world's fastest elevators, travelling 700 ft from the street to the 54th floor in just one minute. Twenty six elevators were required to serve the 30 acres of office space within the building, and the lift shafts took up valuable, rentable floor space (Petroski 1996, p. 200).

The problems of fitting vertical transport systems into increasingly complicated tall structures stretched the capabilities of designers and engineers: one of the most complex solutions was developed in a series of double-decker elevator systems to move people into and up the Eiffel Tower. A number of requirements continued to stimulate technical innovation and better design in elevator systems. These included the importance of speeding up transit times, increasing the throughput of passengers and goods, providing flexibility of use to give individual occupants choice of destinations in vertical journeys and the need for safety with facilities to evacuate occupants quickly in emergencies. Performance improvements were dramatic. By 1931, the 1250 ft tall Empire State Building had elevators which were twice as fast as those in the Woolworth Building. By the mid 1990s, the fastest elevators in the world were installed in Landmark Tower, Yokohama, operating at almost 2400 ft per minute. However, faster elevators created new problems with alignment in shafts, noise, and fears over safety. Some of these problems have been resolved through improvements in roller guides, streamlined and soundproofed cars and advanced ceramic safety brake shoes (Petroski 1996, pp. 200–201). Other technical limitations continue to stimulate innovative responses, such as the use of permanent magnets and linear synchronous motors.

The economics of installing vertical transportation were also important, with the need to limit the space taken up by elevators within tall buildings to maximize rentable floor area and

Fig. 3.6. Petronas Twin Towers sky-bridge

minimize the need to invest in, and maintain, expensive plant and machinery. Around 30 per cent of floor space is taken up by elevators, lobbies and their machine rooms in a typical 100-storey building. Designers and engineers were confronted with the need to optimize the flow of people and goods up and down buildings, particularly at rush hour. Many new designs have been developed, including double-decker elevator cars and multiple elevators working in the same shafts. Many solutions are closely related to the design of buildings. For example, buildings may be divided into a series of stacked, vertical villages, such as the five zones in Norman Foster's Hong Kong Bank (1985). Vertical movement in this building is by elevator to double-height reception spaces for fast commuter traffic: distribution is then by escalator for local traffic within each village zone. Some twin-tower buildings, such as St Lukes Hospital in Tokyo (1994) and Petronas Twin Towers (1996) in Kuala Lumpur (the world's tallest building), use connecting sky-bridges to move people and goods horizontally from one tower to another without the need to descend to the base of the building and join a new elevator complex (Fig. 3.6).

Problems of optimizing internal transportation increase with the size of buildings. For example, the Tokyo Metropolitan Government Building (1992) accommodates 13 000 local government workers who travel through the building on a series of 80 lifts. Buildings such as the twin-towered, 110-storey World Trade Centre (1966–1977) accommodate around 50 000 employees working in 450 businesses and a large number of visitors: the express elevator takes 58 seconds to the 107th floor viewing platform. For the purposes of vertical transportation, each tower is designed as three 30-storey buildings on top of each other. The lobbies of each of these levels are connected to the street by shuttle elevators which travel non-stop to the designated zones, where people change to local elevators. Without the shuttle and local elevator system, the elevator shafts necessary for cars to service all floors would consume almost the total area of the lower floors (Gibson 1998). These complexes often have vast underground concourses with shops and restaurants. Obviously, integrating the movement of people, goods and services within such buildings creates great complexity for designers, space planners, vertical transport and structural engineers.

The technical knowledge necessary to design vertical transport systems includes a thorough understanding of structural *and* mechanical engineering. The integration of structural and mechanical engineering in buildings has remained a prerequisite, unlike in the evolution of mechanical and electrical systems in which there has been increasing specialization divorced from architectural and structural design, often with detrimental consequences to the overall integration of passive and active elements. The structures needed to support vertical transportation systems have themselves often been used to perform a core part of building design. For example, in designs to resist earthquakes, elevator shafts are often used to provide lateral and diagonal cross-bracing. They are at the core of the building and are often its strongest part. It is noteworthy that buildings of 50 storeys and more are generally at higher risk of structural failure from wind loads than they are from earthquakes (Fitzpatrick 1997, p. 169).

3.3. Building controls

Early attempts to automate the control of internal environments focused on temperature and the control of heating systems. The invention of measuring devices and instrumentation were prerequisites for mechanical control. The first major developments in building controls equipment took place in the USA, followed much later by developments in Germany and Switzerland. British engineers generally did not make significant use of control devices until the 1930s.

Crude thermostatic devices, or heat governors, had been in existence since the eighteenth century. Warren S. Johnson patented the first thermostat in 1883; his other inventions spanned automobiles, storage batteries, clocks and telegraphy. While working as a Professor at The Normal School near Milwaukee, Johnson became annoyed at disturbances caused by the janitor's hourly inspections of classroom thermometers, required before heating could be adjusted using dampers controlling hot air circulation from the boiler room. Johnson invented the electric tele-thermoscope in which changes in temperature moved mercury and opened or closed an electric circuit which activated a bell. Thermostats were installed in every room and, rather than disturbing lessons, the janitor responded to bells sounding in the boiler room. The invention became known as the First Johnson System of Temperature Regulation and was installed in schools and other public buildings in the 1890s. The Johnson Electric Service Company was established in 1885. It later became Johnson Controls, now one of the major American suppliers of control systems.

After his initial experiments with combined electro-pneumatic thermostats, Johnson moved to an entirely pneumatic system. These early control systems relied heavily upon pneumatic and hydraulic valves, and compressed air was widely used to actuate equipment. At the same time that Johnson was developing his pneumatic thermostat, Albert Butz developed a crude electrical thermostat. At that time, many smaller buildings and dwellings were not wired for electric power and electrical thermostats were often powered by battery. In 1885 Butz formed the Consolidated Temperature Controlling

Company in Minneapolis. It was renamed the Electric Thermostat Company in 1889 and later became Honeywell Ltd, another major American manufacturer and supplier.

By the late 1920s the first dual thermostat had been invented. This device permitted heating in buildings to be controlled at varying temperatures, with lower temperatures at night and weekends thus improving energy efficiency. As the size of buildings increased, the means of controlling temperature became more complex. Hundreds of monitoring and control devices were installed throughout large buildings, and further attempts were made to centralize control. In the early 1940s this was achieved through the use of capillary tubing, that is liquid-filled tubes providing accurate, but expensive, methods of monitoring temperatures at a distance of up to 50 ft away from the sensing devices. These were superseded by pneumatic control systems with centrally located control panels allowing a single operator to scan room temperatures, ventilating conditions, hot and cold water temperatures and external temperatures around a building.

By 1930 control had moved beyond simple temperature adjustment and theories of control were applied to optimize indoor climate conditions. Electrical control devices were in limited use in Europe: the Rheostatic Company in Britain and Sauter of Switzerland both started producing bimetal thermostats for electric tubular heaters in the mid 1930s. In the 1950s, hydraulic and pneumatic monitoring and control systems were gradually replaced by electrical systems. The first solid state control devices were designed in the early 1960s, and installed in the mid 1960s. Nevertheless, in 1965 about 80 per cent of heating and ventilating systems in the USA were still fitted with pneumatic controls, the remainder being electric or electronic. In Germany only about 20 per cent of such equipment was fitted with central control devices of any type (Billington and Roberts 1982, p. 450). Europe's first computer-controlled air conditioning system was installed at the Paris-Orly Airport in 1971.

3.4. Acoustical engineering

As technologies for heating, lighting, ventilating, conditioning and controlling buildings were increasing in complexity, improvements were also being made in other areas of building performance, including acoustics. Acoustical engineering is based on the physics of sound and is a relatively young discipline compared to other areas of physical engineering such as heating and lighting. Before the Second World War, work by Helmholtz, Olson, the BBC and Sabine were important in establishing the basic principles of acoustics in buildings. Wallace Sabine developed the theory of reverberation time working in the Fogg lecture room at Harvard University. In 1898 he discovered the relationship between absorption and reverberation time, which allowed him to make assumptions about the sound absorption of different materials. This theory led to the ability to build models, simulate sound characteristics and predict acoustics of rooms. The Boston Music Hall (now called the Symphony Hall), which opened in 1900, was the first auditorium to be designed using Sabine's theories. Early testing relied upon the use of gramophone recordings, amplifiers, loudspeakers and microphones, positioned in scale models. Expertise in building acoustics grew rapidly after 1946, but it resulted in many errors as well as some notable successes (Cowell 1997, pp. 124–130). For example, the design of the Royal Festival Hall in London for the 1951 Festival of Britain represented a major step forward in the use of acoustical engineering. Careful, well recorded analysis was made of models simulating the interior of the auditorium. However, there were problems in the calculation of bass response times, which led to further research and the development of Parkin's artificial reverberation system. The use of similar techniques was less successful in the design of the Lincoln Centre in New York in the 1960s.

In the UK, 1:8 scale models were built to help predict the implications of adding or removing absorptive materials. This technique was used, for example, in the Barbican Concert Hall, designed in the 1960s and constructed in the 1970s, and the Olivier Theatre. Acoustic design is stretched to its limits in large concert halls where audiences seek what Steven Groák described as the acoustic equivalent of the telescope and microscope, without amplification (Groák 1992, p. 100). Better models and improved instrumentation for measuring sound helped designers, and by the 1970s better data were available for acoustical engineers and calculations became more accurate. This coincided with the widespread use of anechoic chambers for testing the acoustical characteristics of different materials. In the 1950s and 1960s, research by Amar Bose at Massachusetts Institute of Technology (MIT) resulted in new designs of loudspeakers and an improved understanding of the human perception of sound. The Bose Corporation was founded in 1964 and has since become a major designer and manufacturer of public address audio systems in buildings.

Research on concert halls improved knowledge on cross-reflection of sound and how this affects the perception of quality and richness of music. The theory of early lateral reflection was developed to provide a better understanding of the sense of spaciousness of sound. Further improvements in instrumentation and the use of computer analysis helped improve the accuracy of predictive models. Higher frequency sounds could be measured, enabling smaller scale models to be used, reducing the cost of testing (Cowell 1997, p. 126). By the end of the machine age, knowledge from the design of concert halls was being transferred to the design of other types of buildings. Acoustical engineering was becoming well established as a specialist discipline, but it was the harnessing of digital technologies for acoustical simulation and electro-acoustical tuning of buildings that was to dramatically increase the role and potential of acoustic design in buildings from the 1980s onwards.

3.5. The evolution of integrated mechanical and electrical systems

Thus far, the story of the development of mechanical and electrical services and internal transport systems in buildings during the machine age illustrates three main points about innovation in the built environment. First, a distinct separation in scientific, technical and engineering knowledge emerged between that required for architectural and structural design associated with the passive elements of buildings and structures, and the know-how required to produce the active mechanical and electrical systems installed within them. By contrast, designers of internal transport systems worked more closely with architects and structural engineers. The tension caused by the 'intrusion' of mechanical and electrical systems is vividly described by Wachsmann (1961, p. 106):

> *Entering virtually through back doors and cracks in the masonry, pipes, conduits and ducts of every kind crept along or through the walls, spread over floors and ceilings, pierced threatening holes through beams and columns and thus established a new set of facts completely alien to the original intention.*

The evolution of mechanical and electrical systems for buildings was, nevertheless, similar to that of architecture and structural engineering in that it related closely to the needs of accommodating new social and economic activities.

Second, knowledge about the interaction of one system with another improved as discrete systems were combined to form larger, more complex systems. Third, the integrated nature of interactions between discrete sub-systems such as heating, ventilating, air conditioning and lighting led to further innovations in the control of internal environments. The process involved in the invention of electric light illustrates a particular form of systems innovation — the coordination of energy production, energy transmission, energy metering and lighting. This is also evident in other innovations in mechanical and electrical systems, such as air conditioning, and in the development of internal transportation. It demonstrates the 'systems' or 'complementary' nature of groups of technologies and the ways in which their interactions stimulate the need for further innovation (Rosenberg 1982, pp. 59–60). In the case of electric light, economic success was contingent on considering all aspects of the system in the delivery of light to domestic residences. Improvements in performance of one part of the system were of limited significance without simultaneous improvements in other parts. Changes, however small or apparently insignificant, in the operation or design of one component often influenced the operation or design of other devices. As systems expanded in size and complexity, the efficient operation of the whole therefore required close attention to the interdependencies between component parts.

The demand to provide better control over internal environments increased as buildings became bigger and accommodated a larger number of more sophisticated industrial, commercial and social activities. Mechanical and electrical engineers responded to these demands by developing larger and in many cases more complex technical systems. Technologies were developing at different rates on different fronts. This process is described by Hughes' theory of *reverse salients* (Hughes 1983), in which an expanding large technical system is likened to the unpredictable development of an advancing military line of battle. Referring to concepts taken from military history, a reverse salient is part of the front which lags behind an advancing military front. Applied to technology, a reverse salient is a device or set of components that form part of a subsystem that emerges from the uneven growth of the total system when some components lag behind performance improvements developed in other parts. Thus, inadequacies are exposed and the need for further technical change becomes evident. This has been the case in the evolution of integrated building systems, particularly where centralized environmental control functions have been developed to manage and balance internal temperatures, humidity, and airflow with the heat gains and atmospheric pollutants caused in the use of buildings. It has also been the case in the development of vertical transportation systems. Here, designers and engineers have struggled to optimize the needs of speed, throughput, flexibility and safety with the problems of weight, loss of space and investment and maintenance costs—all in an environment where new materials and mechanical and electrical equipment radically alter calculations.

The task for engineers and others developing new technologies has been to identify in which parts of the system further innovation would be beneficial and then seek technical solutions. To continue to improve performance, innovative activity is directed at removing capacity constraints, or restoring a technical balance and efficient interaction among components in the system as a whole. But the development of mechanical and electrical systems and internal transportation technologies has thus far not been a smooth process. It was often hindered by a lack of knowledge about systems integration because most engineers had been trained to work in one field or another, such as mechanical *or* electrical engineering.

Furthermore, the separation of architectural and structural design capabilities from those needed to engineer mechanical and electrical systems often caused problems for installation as well as for efficient and effective operations within a particular building or structure. Buildings became more complex with the introduction of discrete and then integrated mechanical and electrical systems. This introduced the need to manage in a multi-technology, systems environment. Failure to integrate systems had serious consequences in terms of cost and usability. For example, inappropriate specification of one sub-system, such as electrical supply, could have unforeseen and expensive consequences for the design of other parts of the building, such as ventilation equipment. This is because ventilation ductwork was often designed to cope with assumed heat gains for particular energy usage relating to the designed electric power loads. Over-specified power loads could result in over-designed ventilation equipment which in turn would take up more space, requiring bigger floor to ceiling heights. An unnecessarily expensive building could therefore be produced with high lifecycle costs (Gann *et al.* 1996b).

Finally, buildings in the machine age were often produced for specialized purposes. New specialisms emerged in design and engineering in areas such as geotechnics, structures and mechanics, heating, lighting, electrical engineering and controls, acoustics and vibration. However, innovation in engineering systems often occurred in an environment in which there was a lack of knowledge of systems engineering and where the voice of the user was quiet. In some instances there was public apathy to the potential benefits of new technologies such as those associated with hygiene, health and safety inside buildings.

PART 2

THE DIGITAL AGE

4. Building innovation in the digital age

During the machine age, designers, constructors and materials suppliers were innovative in responding to the needs of producing new types of buildings and structures required by rapidly expanding industrial economies. This was the subject of Part 1 of this book. Part 2 focuses on the contemporary era, dominated by the development of information and communication technologies that are pervading all aspects of society and the economy. This is the digital age, which had its origins in the development of computing in the 1960s and which began to have a major impact on the types of buildings produced and ways in which they were constructed during the 1980s. The main new facilities corresponding to the needs of the digital age are the so-called 'intelligent buildings' and 'wired cities', described in Chapter 5.

This Chapter explores the underlying forces of change in the digital age, including general technical and economic imperatives, emerging markets for intelligent buildings, changes in patterns of property development and ownership, and the use, obsolescence and re-use of different building types. Chapter 5 explains the technical development of intelligent building systems and the ways in which these are produced, arguing that a new paradigm in the production of buildings and structures has emerged with the rise of the digital age. A detailed assessment of this new production system is provided in Chapter 6.

The digital age emerged during a period of economic growth associated with the exploitation of oil reserves and industries deeply rooted in the machine age, such as automobiles, petrochemicals and consumer goods. In the wake of this growth, construction markets became more international, with work expanding in rapidly developing regions such as the newly oil-rich countries of the Middle East and South America. Construction markets also grew rapidly in Japan and South East Asia in response to demand for facilities to produce cars, consumer goods and other products of the machine age.

In the 1960s and 1970s, fixed capital investment associated with manufacturing for the machine age represented a major proportion of construction output. New materials and techniques were developed and traditional practices modernized. The international construction system grew stronger, modifying and utilizing modern industrial techniques, deploying methods of project management first developed in the military and aerospace sectors after the Second World War (Drewer 1990, Linder

1994, Morris 1994). Familiar industrialized construction processes continued to rival traditional craft methods, but both of these were challenged by the emergence of new technologies and organizational principles associated with producing buildings and structures for the digital age.

By the 1980s, another crossroads became apparent in the history of innovation in the built environment because radically different technologies needed to be installed within intelligent buildings to accommodate new types of economic and social activities. At the same time, more modern methods of organizing procurement, design and construction were introduced. These changes affected the internal operations of the construction sector itself, and the wider environment such as the regulatory regimes within which it functioned.

4.1. Rise of the information society

The term 'information and communication technology' is used generically to describe the networks, systems, equipment, software and services developed following the advent of microelectronics. In the early 1970s many of these technologies were in their infancy, but markets for new products and services and supply industries such as computing and software, electronics and telecommunications grew rapidly. By the late 1990s, information and communication technologies had pervaded most areas of public and private life and every sector of the economy in OECD countries. For example, use of the Internet grew rapidly between 1997 and 1999 particularly in the USA and Britain where diffusion into people's homes was possibly more rapid than any other major technology. In Britain, around 11 000 new adult users were logging onto the Internet every day in 1998 and there was a 76 per cent increase in the use of the Internet in people's homes between 1997 and 1998 (Gann *et al.* 1999). The rate of diffusion was much faster than that associated with electrification of homes. The rapid diffusion of information and communication technology in the business sector was mainly due to the large reductions in the cost of storing, processing and transmitting information, coupled with the development of new ranges of products and services capable of revolutionizing production processes in many existing industries (Freeman 1987).

The economic conditions which prevail in the digital age are described by Manuel Castells as informational and global (Castells 1996, p. 66). They are informational because productivity and competitiveness depend upon the capacity to generate, process and apply knowledge-based information. They are global because the activities of production and consumption are organized on a global scale, either directly or through networks of firms and organizations. In the digital age, productivity and competition have shifted away from a focus on manufacturing products towards an emphasis on processes and the provision of services and facilities.

One of the reasons why information and communication technologies have had such a powerful impact on production and consumption is that they can facilitate automation of routine information processing and communication activities. However, they offer more than this, something qualitatively different from advanced automation—these technologies can enable the simultaneous generation and collection of information about underlying productive and administrative processes, providing new levels of transparency to activities that had previously been opaque. When systems are properly configured they may provide feedback, creating new knowledge and a structured approach to learning. This quality which supersedes the traditional logic of automation is what Shoshana Zuboff calls 'informating', offering the capability to transform intra- and inter-organizational processes, and power structures, facilitating new levels of visibility in decision making (Zuboff 1988). For these reasons, the use of information and communication technology has implications for firms and organizations extending beyond the traditional logic of automation of the machine age.

The use of information as a means of control has increased in importance, leading to what Beniger describes as the 'control revolution' (Beniger 1986). The possibilities of providing new types and levels of information are closely related to this and to organizational change and restructuring, or business process re-engineering, fashionable in the late 1980s and early 1990s (Stinchcombe 1990, Scott Morton 1991, Hammer and Champy 1997). Among the leading advocates of the benefits of these changes (such as the supposed paperless office and web lifestyles) are the manufacturers of information systems and services themselves. For example, Bill Gates, Chairman of Microsoft, argues that information technology can enable re-engineering of processes (Gates 1999, pp. 295–329). These utopian views of the benefits of technological change tend to ignore complex social and political interactions through which the technologies are actually developed and applied. The technical and organizational changes associated with information and communication technology have both positive and negative consequences for the organization of work and the ways in which we structure our lives. Such changes can sometimes create fear, mistrust and insecurity among employees, particularly through processes of 'downsizing' when people lose their jobs and sense of community (Sennett 1998).

Both new and old sectors have been affected, including the electronics industries themselves, in which there have been dramatic reductions in production costs and prices of systems.

Time-to-market has become increasingly important for many manufacturers. Digital design and pre-production simulation tools have been successfully deployed to reduce time-to-market and improve quality. For example, between 1987 and 1997, Siemens AG made major improvements in the production of programmable logic controllers used in industrial machine tools. It cut the time to manufacture by a factor of more than 2.5 and reduced the number of defects by a factor of 10 (Gates 1999, pp. 143–144). Production managers in many industries face similar pressures to reduce the time between developing a new product and selling it in the market, particularly in rapidly changing business sectors such as silicon chip fabrication and pharmaceuticals.

In some cases, technical performance in user sectors has been vastly improved. The production of automobiles, for example, has been radically altered through the introduction of new microelectronic process control technologies. Information systems also facilitate closer links between suppliers, producers and customers, thus reducing storage and production time costs. In manufacturing and process industries, evidence suggests that electronics-based production technologies are superior to electromechanical techniques: new microelectronic systems often help to save labour, capital, material and energy. They can also assist in the compression of production time, which itself may reduce risk and improve predictability and quality (Towill 1997). At the same time, the performance of products such as automobiles has been improved through the introduction of computerized engine management systems, global positioning systems, automatic braking and other safety features.

Chris Freeman and Carlota Perez stress the 'systems' nature of information and communication technology which enables closer integration of design, management, production and marketing teams (Freeman and Perez 1988, p. 58). They argue that this facilitates more rapid changes in product and process design and reduces the significance of economies of scale based on dedicated capital-intensive volume-production techniques. Information and communication technology can facilitate the growth of new producer services, together with a reduction in numbers and weight of mechanical components in many products. It is also having a dramatic effect on service industries where market expansion through the provision of new services is occurring, in contrast to the primarily cost-reducing impact on manufacturing. Moreover, information and communication technology can be used to increase the tradability of service activities, in particular those that have previously been constrained by the need for geographical or time proximity of previous patterns of production and consumption. It enables the separation of production from consumption in an increasing number of service activities by bringing in a space or time storage dimension. These technologies can provide the means to accelerate economic and social activities, shortening turnover times and thereby speeding up social processes while reducing the time horizons of meaningful decision making (Harvey 1989, p. 229).

Governments have sought to promote economic growth, competitiveness and improvements in quality of life based on the use of information and communication technology in the digital age. Policies have often been closely linked to the emergence of what is popularly known as the knowledge-based society (Rajan *et al.* 1999). For example, the UK Government's 1998 Competitiveness White Paper spells out changes in scientific, technical and entrepreneurial capabilities it sees as necessary in order to work in the knowledge-based economy (Department of Trade and Industry 1998). Similar reports have been produced by the European Union, the USA, Japan and other governments around the world in attempts to direct policies towards rapidly growing information and knowledge sectors of the economy (for example, see Department of Commerce 1998).

4.2. Buildings in the age of mass communication

These changes are having a number of fundamental consequences on the ways in which buildings and structures are produced and used. New patterns of work, leisure, education, healthcare and travel are emerging, in some cases creating new industries, particularly in the service sector. The diffusion of information systems throughout the service sector—which employs at least 65 per cent of the working population in most advanced economies—is creating the need for a new physical infrastructure. Many service sector employees work in offices, and a growing number have experienced radical changes in working practices. The demand for buildings to accommodate new knowledge workers in areas such as finance and insurance, education and lifelong learning, health and care, design, advertising, marketing and public relations is an important stimulus for change in the ways in which the built environment is designed and constructed. Modern office space has become one of the most important fixed capital investments in the digital age.

The design of digital infrastructures has itself been strongly influenced by lessons learnt from the architecture and design of physical infrastructures such as railways and electrical systems of the machine age. A number of leaders in the development of ideas about the use of digital technologies have themselves trained as architects, for example, Nicholas Negroponte, co-founder of the Massachusetts Institute of Technology (MIT) Media Lab (see Negroponte 1995) and Bill Mitchell, Dean of

Architecture at MIT. Bill Mitchell argues that the Internet has a fundamentally different structure from more traditional physical forms of communication and interaction. It obviates physical geometry, operating under different rules from those that organize action in the public places of traditional cities. For example, an exchange of e-mail can link people at indeterminate locations. By contrast, in the traditional spatially defined city, 'where you are frequently tells who you are and who you are will frequently determine where you are allowed to be' (Mitchell 1995, p. 10). Mitchell argues also that in the digital age, the property industry's traditional cry of:

> *location, location, location becomes bandwidth, bandwidth, bandwidth—tapping directly into a broadband data highway is like being on Main Street . . . the bondage of bandwidth is displacing the tyranny of distance, and a new economy of land use and transportation is emerging . . . in which high bandwidth connectivity is an increasingly crucial variable.*

For many people the world is becoming a smaller, faster, less certain and more complex place. Fixed, well defined spatial arrangements of cities and buildings were traditionally based on specific physical functions and tasks in the machine age, such as sleeping, eating, working, shopping, entertaining, caring and travelling. Time and space were well understood in the heyday of the machine age. There was a time and a place to work, shop and eat, an eight-hour working day and five-day working week. But in the digital age, space and time dimensions are changing, affecting the design and use of buildings. For example, many office buildings, retail outlets, transportation hubs and distribution centres are taking on more flexible forms to accommodate multiple, concurrent and changing activities. Some completely new types of buildings are required, dedicated to the needs of producing and using information technology, services and equipment. These include command, control and switching centres, call centres, multimedia facilities, silicon chip fabrication factories and distribution centres and networks to enable just-in-time (JIT) delivery (Fig. 4.1).

The timescales within which many of these buildings are designed, constructed and used are becoming shorter. In the early 1990s, for example, markets for silicon chips were changing rapidly as successive generations of more powerful chips became available. Product lifecycles were short and time-to-market was critical to business success. Technologies and manufacturing processes also improved, making plant, equipment and factories quickly obsolete. This placed enormous pressures on those designing, building and operating silicon chip fabrication factories—projects often costing over $1 billion.

Fig. 4.1. (clockwise from bottom left) Digital air traffic control centre, Swannick, UK, Tokyo Metropolitan Government Building (exterior and interior views) and the Fuji TV Centre (interior and exterior views), both Tokyo, Japan

Design and construction times were squeezed into little over one year. These factories were expected to pay for themselves and produce a return on investment within two years of operation, before they were overtaken by changes in the market and the next generation of technology. It is difficult to think of examples of fixed capital investment in the machine age on this scale, with such a short payback period.

The changing dimensions of space and time in the digital age sometimes mean that new types of buildings are required in different places. For example, in Japan during the late 1980s there were significant efforts to develop what became known as neighbourhood and resort offices (Gann 1991). The time and space gained by workers in neighbourhood and resort offices is illustrated in Fig. 4.2. Many Japanese government programmes focused on developing local and regional hubs on the new high-speed digital infrastructure around which businesses and communities could cluster. The idea was that neighbourhood offices would reduce the distance and length of time people travelled to work by providing new types of facilities in local plazas, teleports and technopoles within residential areas. These satellite offices were linked to each other and company headquarters using high-speed digital networks. Neighbourhood offices were typically located between 60 and 90 minutes commuting time away from city centre districts. By 1991, an evaluation of six experimental projects had been completed in the Tokyo area. These projects were sponsored by consortia of banks, construction companies, electronics firms and users. The most suitable activities for this new type of working were found

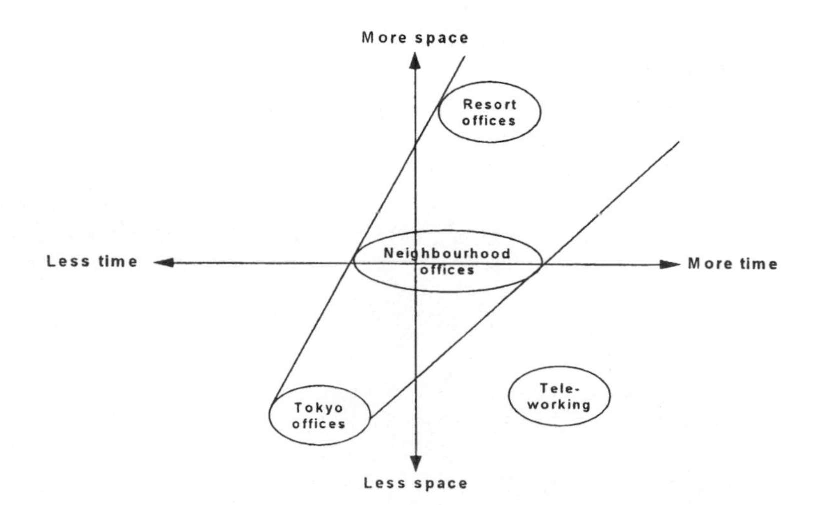

Fig. 4.2. Space and time gained by neighbourhood and resort office workers

to be R&D, design and planning, report writing, computer programming and software design. Employees experienced less stress due to easier journeys to work. Their concentration times had improved and it was hoped that this would result in higher productivity and quality.

Resort offices were the manifestation of a society that wanted to remain connected to work even while on holiday. They provided information networking facilities in remote resorts to enable exhausted or busy executives to re-charge their batteries over a game of golf, whilst keeping in regular contact with colleagues through video conferencing and other communication systems. Pilot studies included evaluation of the impact of natural environments and amenities, such as the effects of humidity, sunlight, the quality of air and water, climate and scenery, upon workers' productivity. Information technology requirements, such as telephone networks, the need for modems, personal computers, teleconferencing, and other remote office facilities, were also assessed. Initial studies suggested that resort offices were suitable for project group working and brainstorming. Thus the digital age had implications on the design of hotels, conference centres and resorts.

More traditional functions continued, but their delivery also changed, often requiring specific types of buildings and facilities designed to accommodate radically different patterns of use. For example, the delivery of medical and health services is changing rapidly with the use of new technologies, such that innovative types of hospitals, doctors' surgeries and health centres are needed. First aid emergency care is increasingly delivered to patients at the point of need, by mobile teams of paramedics. New hospitals are designed around operating theatres and emergency wards in which medical facilities and clean environments with specialized air conditioning and lighting are packed closely together, thus reducing staff and patient movement, speeding up delivery of services. Patients are returned home to convalesce as soon as possible after treatment. Machine age Victorian-style hospitals found in many British towns have become obsolete in terms of layout and facilities. Instead, new medical information networks are needed with local hubs, health centres and specialist units. The production of new facilities to meet society's growing health and care needs is likely to increase in OECD countries during the first quarter of the twenty first century, in parallel with growing numbers of elderly and infirm people. Delivering interactive 'telecare' services into people's homes is becoming possible through the use of remote diagnostics, monitoring and care systems. Achieving this will require housing designed and built to accommodate new digital infrastructure and facilities (Tang *et al.* 2000). Such changes raise many ethical questions about the use of technology, including concerns about telesurveillance and loss of independence and control. The introduction of technologies requiring registration for access and use, such as security cards in buildings, is generating widespread debate about the protection of privacy in modern society (Abercrombie *et al.* 1986, pp. 148–152). However, such technologies also offer benefits of more effective and efficient communication, for example between those who need assistance and those who can provide it. The pace of change in the built environment is often largely determined by the rate at which issues concerning use are resolved, including the need to revise regulations and understand costs and benefits.

In the digital age the challenge is to design and produce buildings that cater for multiple and concurrent patterns of use made possible by information and communication technology. Mitchell (1995) argues that the cultural values and symbols expressed in buildings and structures are taking on new meanings as traditional physical notions of space and time decompose and are replaced by what he calls reconfigurable, asynchronous software constructions. Table 4.1 illustrates some of the ways in which digital technologies are beginning to replace or augment more traditional physical functions and places. In the machine age it was usually possible to determine the use of a building by examining its floor plan or footprint (Pevsner's *History of Building Types* (1976) can be used to explore this relationship). However, the demand for more flexible, multifunctional spaces for different types of information and knowledge-based activities in the digital age is beginning to change this. It is no longer so easy to learn of the physical use of a modern adaptable building from its footprint or floor plan.

The demands of the information society are changing the conditions of space and time in the built environment. No longer are these concepts commensurate with the mechanized world described by Sigfried Giedion in his classic book, *Space, Time and Architecture* (1967). Space and time are simultaneously becoming compressed and perhaps more flexible concepts because of the effects of the use of information and communication technologies and new patterns of social and economic activity (Graham and Marvin 1996). The speed up of lifecycles, use-cycles, design time, construction time, maintenance time, adaptation time, all relate to increasing pressures on the ways in which we use space—what Karl Marx called 'the annihilation of space by time'. Equally, more and more activities occur under pressures of shorter time spans. The need to make a quick return on investment changes the nature of fixed capital formation in buildings and structures. Utilization rates and shorter building lifecycles change the ways in which space is used.

Table 4.1. From the machine age to the digital age (source: adapted and extended from Mitchell (1995))

Machine age	Digital age
Bookstores	Bitstores
Stacks	Servers
Galleries	Virtual museums
Theatres	Entertainment infrastructures
Universities	Virtual campuses
Hospitals	Telemedicine
Prisons	Electronic supervision programmes
High street banks	Automatic teller machines
Trading floors	Electronic trading systems
Department stores	Electronic shopping malls
Work	Net-work
At home	@home
Offices	Tele-cottages
Business centre	Teleworking centre
Real estate	Cyberspace
Wild West	Electronic frontier
Face-to-face	Interface
On the spot	On the net
Street networks	World Wide Web
Enclosure	Encryption
Public space	Public access
Territory	Topology
Cities	Technopoles
Cafes	Internet cafes
Building envelopes	Smartskins
Heating control	Environmental and energy management systems

The emerging networked society no longer requires the old fixed certainties of time and place because networked relationships can be clustered across indefinite ranges of time and space (Castells 1996, Chapter 6). Castells argues that the new spatial order is one of flows instead of one of places. In some cases, places and their buildings are no longer required at all. Mitchell suggests that software is increasingly beating hardware. He cites the case of Columbia University which, in the early 1990s, scrapped plans to build a $20m addition to its Law Library and instead bought a supercomputer and embarked on a programme of scanning and storing tens of thousands of deteriorating books (Mitchell 1995, p. 48). The storage of data in the form of bits replaced the storage of physical artefacts in the form of books. All that was solid had melted into air (Berman 1982). The ability to minimize the space taken for storage reduces the need for fixed storage places. At Columbia, this form of technical change cost architects and constructors the work of building a physical library, whilst the computing and communications industries won new orders for a networked supercomputer. Moreover, information held in the new virtual library could be made available to people when and where they wanted it, anywhere in the world. The traditional pressures of space and time to visit the physical location were removed.

Similar examples are evident in the case of electronic banking. These examples serve to illustrate the substitution of plant, equipment and now software for physical bricks and mortar, lending weight to the arguments made in Chapter 1. It serves to illustrate the changing physical, spatial and temporal nature of the built environment in the digital age, in which practices and places are being transformed. Fitting changing working practices into buildings and designing buildings to accommodate new types of activities have resulted in a form of spatial and organizational re-engineering, sometimes called 'process architecture' (Duffy 1997, Horgen *et al.* 1999).

4.3. Britain in the 1980s — a new generation of office buildings

The rise of the information society brought with it huge demands for new types of office space. As a consequence, the property and construction boom in Britain during the 1980s resulted in the biggest transformation of the landscape of the City of London since the Second World War. In 1988, the total office stock was estimated to be around 6.5 million square metres, with a further 2.8 million square metres under construction in and around the City of London (Debenham *et al.* 1992) and planning permission for an additional 1.8 million square metres. The London office stock was to increase by around 25 per cent between 1986 and 1992. An idea of the scale of this development is provided through comparison with previous periods of construction. For example, it took 40 years for the redevelopment of 20 per cent of the buildings in the City in the period between 1900 and 1940 (Duffy and Henney 1989). Only about 900 000 square metres of office space were constructed in the previous boom in the early 1970s (data from Barras (various dates) and *Financial Times*, 18 September 1992). London had re-invented itself with a shockwave of building activity stretching from Stockley Park in the west to Canary Wharf in the east, creating some of the best new office accommodation in the world.

The growth of information-intensive work initially had the greatest impact on business districts in city centres, concentrated in London and the south east of England. In 1988, around 75 per cent of all employment in the City of London was in office work. In Greater London, employment in financial services grew by 200 000, from around 520 000 in 1981 to 720 000 by 1986 (Duffy and Henney 1989). In 1986, firms in the City of London invested in excess of £1 billion in equipment such as dealer boards and desks, new telecommunications systems and other related information technology and back-up services (estimates from PA Consultants). By 1987, it was estimated that banks spent a further £2 billion on information technology, in addition to another £1 billion invested by other institutions in the City (Duffy and Henney 1989, p. 33). Substantial sums were spent on word processors, fax machines, mainframe and personal computers. This new machinery meant that electric power consumption in the City increased by one third between 1985 and 1990. The need to cool offices because of the extra heat generated by so much electrical and electronic equipment resulted in cooling loads increasing from an average of 15–20 W/m^2 to 100 W/m^2. New buildings were needed to accommodate all this new technology. The 1980s construction boom was therefore driven by new space needs and building technology requirements in offices – particularly in the financial services sector. Buildings to house electronic dealer rooms for international securities markets were in the vanguard.

Financial and business services firms were early adopters of information and communication technology. They were also innovators, developing new services which led to different working patterns. There were strong pressures to improve productivity and gain better access to markets through the use of new information systems in many service sector activities, including office and retail work. In order to accommodate the changes brought about by information-intensive work, firms needed to adapt existing buildings or move to new ones.

New internal information systems needed to be connected to local and long-distance communication networks in order to gain the full benefits of faster and better links with customers and suppliers. By the mid 1980s, large multinational corporations were demanding buildings to accommodate network management and control centres for their global telecommunication networks, as well as electronic office equipment, cabling and space for new desks used by information workers. In many ways these buildings were analogous to stations on railway networks or shipping ports for commerce, the difference being that data were transmitted instead of passengers and goods. Multimedia hubs lay at the heart of many new commercial buildings, facilitating local access to long-distance communications networks. They were used as multiple access points interconnecting wide, metropolitan and local area networks. Different types of cabling, such as twisted pair, coaxial or optical fibre, and different local area network architectures needed to be integrated to distribute information throughout these buildings. Accommodating such facilities at one location and distributing information in the local area often required the construction of intelligent building complexes.

The use and control of information became of central strategic importance to organizations whose demands for new technology in buildings increased rapidly. The rapid rate of change in information and communication technologies themselves made planning for future needs difficult. Building users therefore demanded adaptable and upgradable cabling for data and communication networks and variable air conditioning and environmental control systems to remove excess heat generated by office equipment. Moreover, increased reliance on technology brought new demands for systems that could avert disaster should failures occur. These included clean, uninterruptible power supplies, fire protection and security systems. Finally, users wanted adaptable space to accommodate changing organizational structures. The ability to reorganize the use of space became a critical element of competitive success for many firms who needed to respond to rapidly changing market conditions. Large continuous and adaptable floor spaces became fashionable, replacing the rigid layouts of 1960s and 1970s cellular offices (Fig. 4.3). Evidence of the advantages of having adaptable space could be seen in the ways in which some firms were able to swiftly move their operations into new buildings after terrorist bomb attacks destroyed office buildings in London in 1992 and 1993.

Fig. 4.3. A modern, open plan dealer room – SBC Warburg (1998)

The most successful new buildings were those designed to facilitate changes in use through the inclusion of movable partitioning, furniture and scenery and the use of appropriate cabling architectures. Nearly 80 per cent of companies moving into new buildings changed their floor layouts within the first twelve months of occupancy. Savings could be considerable if buildings could be easily adapted. In the case of BP Oil's new Hemel Hempstead office building, completed in 1990, a person could be moved to a new desk location in just ten minutes, compared with a time of at least one day in the old London headquarters. Similar savings were made by Chase Manhattan Bank's new headquarters in Bournemouth, where a move could take less than two hours compared with up to one week in the previous building.

The potential benefits of using new information and communication technologies in buildings were not confined to the financial services sector or to corporate information technology departments and headquarters. Large retailers were searching for ways to reduce running costs such as energy consumption and to improve inventory and trading systems using EPOS (electronic point of sale) or EDI (electronic data interchange). Work in research, development and design departments and in high-technology businesses such as microelectronics, software, pharmaceuticals and new materials was becoming more information intensive. Increased use of information technology changed space utilization considerably. By 1988, computer and electronics firms themselves used about 20 per cent of their floor area for computer facilities. In new offices, computer systems took up six per cent of floor space, at the same time as the area of floor space per person decreased. Similar trends were occurring in factories. In the 1960s very little factory floor area was allocated for office and information processing purposes; by the 1990s this had increased dramatically to over half the available area in many cases. The quantity of information technology and mechanical and electrical equipment used by large retail chain stores also increased considerably through the 1980s. One indication of the increase in use of these new systems is the cost of mechanical and electrical installations as a proportion of total construction costs, discussed in Chapter 5. This evidence gives further weight to the argument about the substitution of plant and equipment for buildings and structures in patterns of fixed capital investment.

Nevertheless, in the early 1980s, equipment was initially too expensive and returns on investment too uncertain for most users of small and medium sized buildings to consider. However, diffusion occurred rapidly as costs came down and systems became easier to install and use. By the 1990s, systems were installed in a wider and wider range of buildings, including public sector offices, defence establishments, universities, schools and hospitals. Remote monitoring and control facilities meant that applications were not confined to large properties. For example, the use of building control technologies became widespread, particularly in small-scale installations in branches of banks and building societies, retail chains, public houses and schools. These were networked to central control locations, facilitating monitoring and control of energy consumption, fire protection, security, lighting and refrigeration. In these cases, building owners achieved benefits from linking together a number of smaller buildings.

4.3.1. Changing property markets

The form of property ownership and type of user had a direct bearing on the kind of buildings constructed and the ways in which they were built: the existing structure of property provision posed a major constraint to the development and use of new information and communication systems. The commercial property sector worked as a series of linked markets, bringing together property users, investors and developers (Nabarro 1990). The user market comprised owner–occupier firms buying and selling office space. The property investment market was made up of specialist property companies and financial institutions which owned and traded buildings, the aim being to secure a return on investments from rising rental and capital values. The property development market was where new buildings were produced and old buildings redeveloped or scrapped.

Office buildings were either constructed for specific owner–occupiers, or built speculatively by property developers seeking a return on investment by selling on, or leasing to tenants. The existing structure of property acquisition was based on long-term investment and returns on long leases, typically of 25 years. Rental property was seen as a fixed cost for long-term investment in which upward rent reviews were written into contracts. Tenants were locked into a system in which there was little flexibility over the terms and conditions of leases. Landlords had little understanding of how tenants used their buildings and paid scant attention to changing user requirements. This was the 'let-and-forget' system in which institutional investors were only concerned about the profit margins on long leases. Property owners had no idea whether their properties were operating efficiently or providing users with the facilities they required.

Under these ownership structures it was perhaps not surprising that many of the innovative buildings designed to accommodate information-intensive work were constructed specifically for multinational corporations, such as the Chase Manhattan Bank (Bournemouth), Rank Xerox (Marlow) and the Lloyds Building (London). But a new generation of speculatively developed property also emerged, designed to accommodate the needs of modern information users. These offices included huge developments at Broadgate, Canary Wharf, and the Isle of Dogs, Stockley Park, and many others in and around London. Both owner-occupied and speculative developments relied upon the design and installation of bespoke systems for communications and control of internal environments. However, during the 1980s, buildings constructed for owner–occupiers were more likely to meet users' needs than speculative developments where end-users were unknown until after initial designs had been completed.

The huge scale of the 1980s property boom in Britain was made possible by a number of structural changes in traditional property institutions, such as restructuring of the property development industry, new methods of financing projects and new planning legislation. Two types of property developer were active — investor developers and trader developers. Large investor developers such as Land Securities or MEPC tended to be cautious about new technologies and innovative designs, placing more emphasis on managing their portfolios than on developing new buildings. They typically developed new buildings to enhance portfolios and funds were often raised from internal revenue streams, thus ensuring some degree of protection from insolvency caused by high debts during recessions.

By contrast, trader developers were entrepreneurial, aiming to build and sell property. Most were very small companies with perhaps two or three partners and a small staff. Their work involved spotting opportunities for property development and bringing together the necessary participants from planning, finance, construction and estate agency to construct, let and sell buildings. Their critical skills were associated with finding new funding mechanisms.

In general, neither investor developers nor trader developers had a particularly good knowledge of how to manage construction processes, although trader developers often had close working relationships with architects. This was reflected in the 'visionary' nature of many of the buildings commissioned by flamboyant trader developers such as Stuart Lipton of Stanhope. Lack of knowledge about production made it difficult to integrate new technologies into buildings during the 1980s. Trader developers had little to do with end-user markets and traditionally obtained their market intelligence from chartered surveyors and letting agents: weak performance by both led many trader developers to turn to other more reliable sources for market intelligence, such as independent property market consultants. However, they faced tensions between meeting the different requirements of their two customers: investors who expected a return and building occupants who expected buildings to meet their changing needs. On the one hand they had to build cheaply enough to maximize returns and on the other, they needed to install more facilities and create a space capable of supporting a variety of different and changing activities. Under these circumstances, it was difficult to develop innovative buildings to accommodate the needs of information workers.

Many developers failed to balance these pressures, often with the result of 'locking out' potential tenants because insufficient space was left to accommodate equipment, cabling and air conditioning, or, occasionally, because the space was over-specified and too expensive. In an attempt to overcome some of the difficulties of designing buildings for unknown users, developers adopted a North American approach called 'shell and core' construction. This involved constructing the basic structure and envelope and installing the minimum required in terms of general services such as lifts, bathrooms and toilets, fire protection and security systems. Developers provided space in the form of raised floors, suspended ceilings and riser ducts, in which tenants could 'fit out' open plan floor areas to their own specifications. This resulted in what was in effect the construction of buildings within buildings in which developers provided the basic infrastructure and supports, and tenants installed the interiors. This approach echoed ideas relating to levels of permanence, change and building lifecycles championed by the Dutch architect Nicholas Habraken in the 1950s and 1960s (Habraken 1972, 1998).

4.3.2. New forms of finance

The second institutional change in property development was associated with new ways of financing projects. Patterns of property investment in Britain differed from those of the 1970s. In the early 1980s there was a slowdown of institutional investment in property as leading investors restructured their portfolios to benefit from new sources of investment, particularly in equities. Figure 4.4 illustrates the total returns on investment from property compared with equities and gilts between 1981 and 1989 in the UK; Fig. 4.5 shows the change in capital values of property relative to these other forms of investment during the same period.

A funding gap emerged between property users and investors. The funding gap meant that alternative financial instruments had to be found to fund property development. Loan financing grew rapidly, especially by overseas banks, using highly complex instruments. The gap was also filled by new, specialist trader developers, such as Stanhope, Rosehaugh and Mountleigh. Much higher levels of trading in property were experienced compared with the 1970s, as fund managers realigned their portfolios to dispose of old property and reinvest in new, more modern buildings (Key *et al.* 1990).

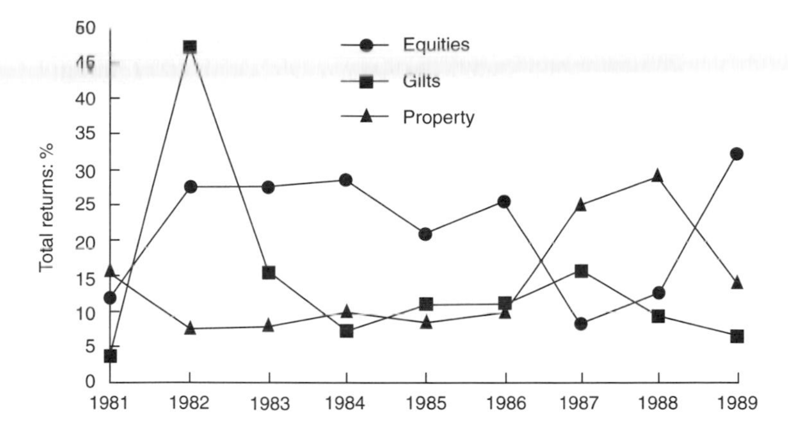

Fig. 4.4. UK property returns compared with equities and gilts, 1981–1989 (source: IPD and BZW (Nabarro 1990, p. 57))

Cycles in the property development markets had been typified by changes in the balance between investor developers, such as large institutional pension funds, and new, highly geared speculative trader developers. There was usually a shake-out at the end of each boom in which many trader developers went into liquidation. A few survived and maintained a portfolio of properties eventually to become investor developers. Deregulation of financial markets had a dual effect on property markets and the structure of ownership and development. Not only did it result in investment in information and communication equipment and networks by the financial services sector itself, it also led to new flows of capital into the City in equity markets and

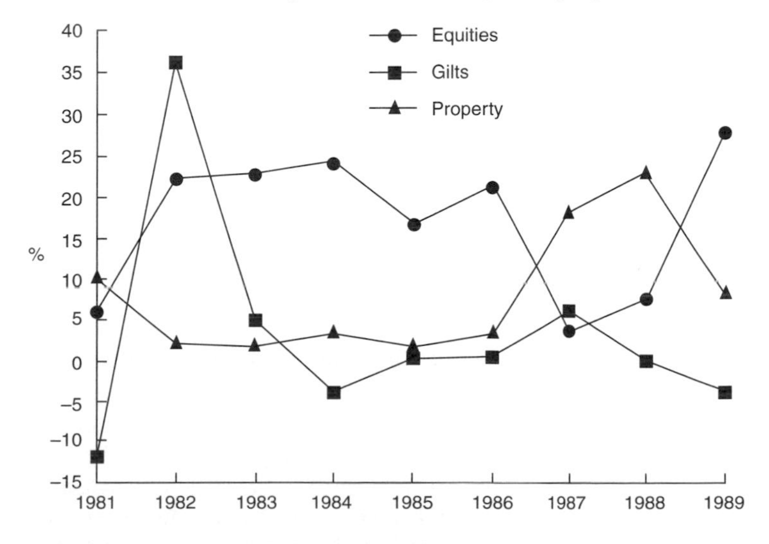

Fig. 4.5. Change in capital values of UK property compared with equities and gilts, 1981–1989 (source: IPD and BZW (Nabarro 1990, p. 57))

property investment, which helped to fund property development. Between 1987 and 1990 around one third of capital investments came from overseas (Daniels and Bobe 1990). As well as attracting international finance, the British property boom was marked by the involvement of several international trader developers, the most famous being the Canadian firm, Olympia and York. These international players were partly responsible for stimulating the use of new approaches such as shell and core construction.

4.3.3. Deregulation

The third area of change which helped to stimulate the property boom was related to regulations governing land use and development. In Britain, the planning system traditionally lags behind social and economic change, and planning control often acts as an institutional constraint on technological change. For example, it was largely due to constraints imposed by height restrictions under the London Building Acts that high-rise office buildings were not constructed in Britain until the 1950s (Atkinson 1993, p. 133). Legislation affecting the location, design and use of buildings tends to follow rather than lead events. Government action and the regulatory procedure are invariably based on an existing legal and administrative framework. In the 1980s, pent-up demand for new types of buildings was partly released by revisions to 1970s planning restrictions, for example, when plot ratios (the ratio of developed space to plot size) were increased from 3.5:1 to around 5:1. The growth of service sector activities together with the use of new information technologies and production processes in light manufacturing resulted in the blurring of boundaries between office and light industrial activities. Premises began to be used for a mixture of different activities as work patterns and industrial output changed. Property developers and building owners responded by seeking planning approval for new types of multipurpose, flexible buildings and by putting pressure on government to change what had become an out-of-date planning system.

A process of deregulation ensued in which a new B1 Business Class Use Order was introduced in the Town and Country Planning (Use Classes) Order 1987, together with revisions in the General Development Order 1988. For example, normal planning controls were relaxed in special Enterprise Zones and Urban Development Areas, such as London's Docklands. The new B1 Use Classes brought together three categories from the previous 1972 Order: these were offices, R&D laboratories and studios, and light industry. This made possible the development of 'high-tech' out-of-town business and science parks, with space for parking and room to extend buildings. The specific aim was to reduce the intervention of planning in the use of commercial property taking account of the need for more flexibility in the use of space. As a result there were benefits to users and property owners through the improvement of economic and functional efficiency with which space was used. However it is likely that the new Use Classes also introduced a number of adverse effects such as an increase in road traffic and confusion over the correct interpretation of the 'amenity test' (Wootton Jeffreys Consultants Ltd and Thorpe 1991).

4.4. Experience in other countries in the 1980s

4.4.1. Europe

During the 1980s, Britain appeared to be more advanced than most other European countries in the development of new types of buildings for the digital age. Estimates of the proportion of office floor space suitable for information workers suggest that in 1989, London was the most advanced with around 12 per cent, followed by Paris with 10 per cent and Frankfurt and Madrid with 5 per cent. This position was partly achieved because of the types of firms occupying property in Britain and their use of advanced telecommunication systems. Britain's status as a 'transatlantic hub' meant that it was a location favoured by American and other multinationals operating in Europe (Sauvant 1986, Morgan and Sayer 1988). The City of London probably had the largest community of advanced information service business users in Western Europe and in the late 1980s, Britain had more high-speed leased telecommunication lines in use than any other European country. During the 1980s, office users in France and Germany tended to have less sophisticated demands for technologies in buildings than their British counterparts. For example, major banks in Continental Europe were slower to introduce new information and digital trading systems than their counterparts in Britain.

There was a marked contrast between the approaches taken in the development of buildings for the digital age in Britain and those in other North European countries, notably Scandinavia, the Netherlands and Germany. The British approach was in many respects similar to that developed in North America, heavily biased towards the supply side and the commercial interests of property developers. In contrast, user influence on office design was much higher in Scandinavia and the Netherlands, where the best offices provided higher standards of space, amenity and comfort compared with those achieved in Britain or the USA. The North European approach related office design more closely to achieving organizational goals, and buildings were viewed as a direct corporate investment (Duffy *et al.* 1993, pp. 8–9).

European markets for new digital age buildings differed in other respects from those in the USA and Japan. Prior to the 1980s development boom in Britain, the size and proportion of new building activity in the USA and Japan had been much greater than in Europe. As a consequence, design and systems development was geared more towards technologies for installation in new buildings. The new communication and control technologies were therefore not necessarily compatible with the needs of upgrading or retrofitting into existing buildings. They were not always appropriate for use in Europe where the replacement rate of buildings had been lower. Nevertheless, developments in the USA in the early and mid 1980s exerted a strong influence on the type of products installed in Europe, particularly in Britain, where many US suppliers entered the market.

4.4.2. The USA

The intelligent building concept first emerged in the USA in the early 1980s. Development of building information technologies advanced there more rapidly than elsewhere. This was partly because of the different approach to property acquisition and the greater extent to which telecommunications was liberalized. A wide variety of telecommunication services were available to North American users: in the late 1980s the USA had a 43 per cent share of the world market for information technology services, Western Europe accounted for 29 per cent and Japan for just 18 per cent. American landlords and building managers were legally permitted to provide network services to building users. In the early 1980s new firms were established to exploit what they perceived to be market opportunities by linking building management with shared tenant services (STS). Third party providers installed and maintained centralized telephone services within buildings. Building owners, together with STS providers became in-house utility companies, offering lessees cheaper value-added services. The promise from STS providers was that the telephone-based systems could effectively control building information systems. The idea was that building occupiers should be able to move into new spaces and plug into terminals connected to networks managed by landlords. Telecommunication firms, landlords and property developers 'pushed' these services in big marketing drives. Thus the digital age brought with it a marked shift from buildings as physical places to facilities offering serviced spaces.

The trend in technologies in the USA was towards the provision of multi-channel intelligent networks. Different services were usually provided down single communications lines linked through mainframe computers. Each suite reported back to a mainframe which recorded usage. System faults were deemed to be the landlord's responsibility. Occupiers paid a fixed rate for unlimited use and it was therefore in the landlords' interests to encourage activities such as energy saving, unlike the British experience of multiple occupancy buildings in which tenants looked after themselves.

Large, multi-tenanted buildings were common in the USA. Building space was rented gross, including all taxes and charges for utilities. Shell and core methods of construction were widely practised. This involved the provision of main vertical risers and connections to local and long-distance network suppliers and satellite communications by landlords. Tenants fitted out their space by 'patching' into risers. Cables were threaded through hollow steel decks, or taken within suspended plenum ceilings. Landlords had the right to install cables and could disturb neighbouring tenants in order to do so. For this reason, the use of raised flooring was not as widespread in the USA as it was in Britain, where access had to be gained from within the same tenanted space. American landlords appeared to be better at managing facilities in buildings than their British counterparts. They generally took responsibility for all mechanical and electrical maintenance and associated activities from simple functions such as changing light bulbs, to managing complex space planning issues. In large American cities, tall buildings accommodated hundreds of different firms with continually changing space requirements. Management of space reallocation required sophisticated techniques that were often lacking in Britain. By contrast, British landlords practised the 'let-and-forget' approach, offering long leases and often leaving the management of services and maintenance to tenants.

By the late 1980s, American concepts of servicing buildings for information workers were changing. Shared tenant service providers had failed to meet users' changing needs and the push by telecommunication vendors failed to materialize in the form of large new markets. Emphasis shifted towards the systems offered by environmental control companies, consulting engineers and new electronics firms all of whom were offering digitally controlled building information systems. Furthermore, the nature of property development hindered attempts to install sophisticated information systems within buildings. The speculative construction of properties for quick financial return meant that there was little incentive on the part of developers to invest in expensive systems. New buildings were designed for short lifecycles and there was often little point in increasing costs by installing expensive equipment. For example, in New York, offices could be written off after five years, while new business developments around London had a 25-year design life instead of the traditional 60 years. These issues imposed new constraints to the introduction of technology in buildings, made worse by the slow rate at which architects and the design community embraced the need to accommodate new information and communication technologies. After an early lead, the USA appeared to be overtaken by developments in buildings designed to facilitate work with information systems in Europe and Japan. In the USA there was no clear pattern of development, little government sponsorship (as in Japan) and few industry standards to help establish a coherent market for high-quality office environments. The effects of short-term property speculation were evident in the many low-grade buildings in Silicon Valley, at the heart of the digital revolution. Pressure to develop fast and build cheap created fierce competition between developers resulting in cost reduction rather than improvements in office quality. The exception was when users, such as some large software companies or international banking organizations, had direct involvement in specifying their needs and in installing and operating technologies in bespoke owner-occupied buildings.

4.4.3. Japan

The context in which buildings for the digital age were developed and used in Japan was very different from that in the USA or Europe. In particular, problems caused by rapid urbanization led to congested cities and cramped office space. The growth of large cities in Japan was as dramatic as its rapid economic rise: cities grew quickly during the 1950s and 1960s, with little planning or coordination. In the 20 years between 1960 and 1980, the area of densely inhabited districts increased 2.6 times and the population of these districts 1.7 times. By the 1980s, suitable land for building new facilities was scarce in large, cramped cities such as Tokyo and Osaka. Most Japanese cities were crowded, transportation congested, and in many ways the infrastructure was over-stretched and under-developed. Land costs were very high, although they began to fall in 1991, for the first time in 16 years. However, urbanization continued through the 1980s and it was predicted that 70 per cent of the total Japanese population would live in cities by the beginning of the twenty first century.

Building for the digital age in Japan has been made more complicated because many cities are located on major seismic fault lines. The need to construct buildings capable of withstanding major earthquakes has influenced decisions to install elaborate disaster-prevention technologies. By the 1980s, these included the use of digital communication and control systems. Other differences include the role played by the Japanese government in promoting the development of construction and information technology industries. End-users were also more involved in the development of their buildings and facilities than in most European countries or America. For example, most Japanese intelligent buildings were constructed for owner–occupiers whose communication and data handling needs had grown rapidly. Many of these were large multinational corporations.

Japanese offices were traditionally open plan, long and narrow, to allow natural lighting. Open plan work areas permitted face-to-face communication typical of the more 'cooperative' style of work patterns found in Japan. The history of Japanese office work differs from that in the West. In particular, there was no progression from the use of typewriters to the use of desktop word processors, because in Japan, the use of typewriters never became widespread. The complexity of Kanji—the Japanese script—meant that one highly skilled typist would often serve as many as 100 office workers. Typists were easily accessible because of open plan layouts. The introduction of desktop computing was sudden, occurring from the mid 1980s onwards, and had dramatic implications for space needs. Existing offices did not have enough room for the additional hardware and there was large pent-up demand for new modern office facilities. Between 1981 and 1985, approximately 1.1 million square metres of office space was constructed per annum in Tokyo. Almost triple this amount was constructed per year between 1985 and 1990, although much of this space was for standard offices and did not include facilities for working with modern information and communication systems.

The intelligent building concept was first used in Japan around the mid 1980s; by the late 1980s, first generation showcase buildings had been constructed. Most were large, expensive, owner-occupied buildings incorporating bespoke, experimental systems. The development of these digital age buildings coincided with a period of urban growth and infrastructural modernization on a huge scale. The Japanese economy was shifting away from reliance on energy- and labour-intensive industries such as shipbuilding and textiles, towards high value-added industries such as semiconductors and computers. Industry and government believed that the construction of intelligent buildings linked to new information networks would be the infrastructure at the heart of the economy in the 1990s. An increasing number of firms were embracing major organizational changes, with radical implications for their space needs. As a consequence the demand for intelligent buildings grew. The main users of these buildings were large multinational financial services, electronics, telecommunications and motor vehicles corporations who built prestigious headquarters, for example, NTT's Shinagawa Twins, the NEC Supertower (Fig. 4.6), and Kajima's KI building.

There was also a trend towards the provision of tenanted buildings with STS, particularly after the deregulation of telecommunications in 1985. Several speculative intelligent buildings were constructed, such as the Osaka Crystal Tower, the Shinjuku L Tower and the Ark Hills and Shiroyama Hills complexes in Tokyo. However, Japanese designers and contractors failed to resolve the issue of how to meet space and equipment needs for different users in tenanted buildings. Their

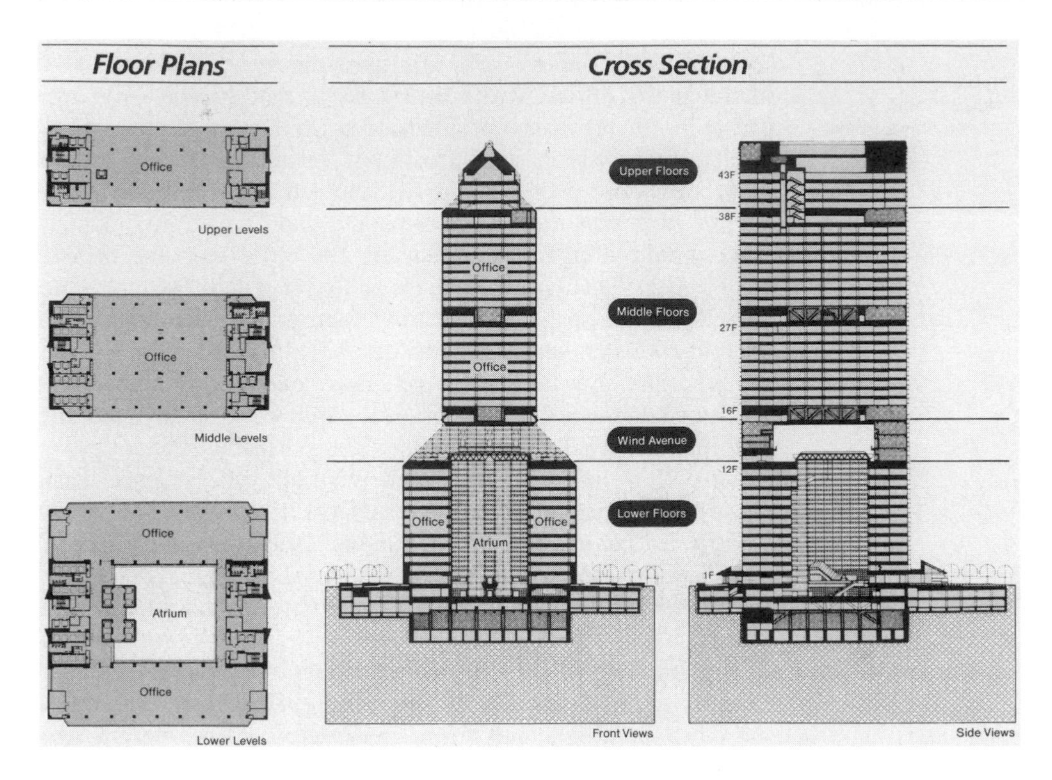

Fig. 4.6. NEC Supertower

approach, however, differed from that in the West in that they tended to install more services initially.

By the late 1980s, national and local government also constructed incredibly expensive and technically sophisticated intelligent buildings, for example, the Patent Office Building, the Tokyo Metropolitan Government Building and, in the 1990s, the Tokyo International Forum (Fig. 4.7). In other cases new complexes were constructed through public–private sector joint ventures such as Sonic City in Saitama Prefecture.

In the 1980s, the Japanese economy was based on rapid innovation, in which the development of information technology played a prominent part. The Japanese people bought a vast range and number of electronic goods and the population was probably more computer literate than in most other countries. Information technology industries were growing quickly. Developments aimed at the widespread use of information technology fell into four categories: intelligent cities, intelligent buildings, intelligent homes and smaller equipment sub-systems. A variety of government programmes had been introduced to facilitate technology-intensive production and to provide international producer services, such as finance and research. Six major development programmes sponsored by four ministries stimulated the potential to construct intelligent buildings. These were

Fig. 4.7. Tokyo International Forum

Technopolis (Ministry for International Trade and Industry, MITI), Research Core (MITI), New Media Community (MITI), Teletopia (Ministry of Post and Telecommunications, MPT), Intelligent City Programme (Ministry of Construction, MoC) and Greentopia (Ministry of Agriculture, Forestry and Fisheries, MAFF).

Intelligent buildings were perceived by users to play as important a role in improving the efficiency of office workers as automation played on the shopfloors of manufacturing industries. The need for new, highly serviced office space was recognized as crucial to enable firms to reorganize their

Fig. 4.8. Kansai airport, built on an artificial island in an earthquake zone

operations so as to become more innovative. This had a marked impact on the location, type, size and use of buildings, and the technologies installed within them. Space was scarce and improving the built environment to meet these needs meant building up, down and out in high-rise, underground and ocean cities (Fig. 4.8).

The construction of new business districts, called 'intelligent plazas', proliferated in large Japanese cities. New buildings functioned as central and relay stations at the node points of new communication networks. The idea was that they should become the access points for information from dealing rooms and for shared supercomputers, technological, business and administrative databases, video conferencing rooms, and optical fibre, satellite and microwave communication networks. A wave of optimism resulted in the MoC expecting more than $300 billion (1990 prices) to be invested in these buildings during the 1990s; over half of this amount was thought to be needed in cabling, electronic equipment and services. The MoC offered a subsidy in the form of a loan attracting lower rates of interest to encourage the development of small scale intelligent buildings. It claimed that it was essential for inter-city telecommunications networks to be set up concurrently with and inseparably from land redevelopment and new urban renewal projects (Gann 1992). By 1991, at least 1000 regional redevelopment projects had been sponsored, at the core of which involved the construction of intelligent buildings.

The pace of development in Japan was being driven by four particular features of economic restructuring: liberalization of telecommunications; deregulation of financial services;

increasing demands by large corporations for swift national and global communication facilities; and government sponsorship and promotion of information technology industries. The Japanese government regarded intelligent buildings as central to the nation's infrastructure and development of the information economy. It therefore played a more prominent role in the promotion of digital technologies in buildings than the British or American governments, at that time.

Energy conservation was also a major driving force behind the inclusion of new control technologies within buildings in Japan. The oil crises of 1973 and 1979 were particularly severe in Japan. Efforts were subsequently made to conserve energy and to substitute other energy sources for oil. These efforts were coordinated by MITI with dramatic results. Energy consumption was reduced per unit of output faster than in any other OECD country, partly as a result of energy conservation and partly as a result of structural change in the economy with less emphasis on heavy, energy-intensive industries. Energy savings were higher in new intelligent buildings. For example, NTT's Shinagawa Twins building achieved savings of 17.5 per cent over an equivalent conventional building. This gave a simple payback of five years on the additional capital cost. The NEC Supertower used 41 per cent less energy than its conventional equivalent, and similar performance was achieved in the Shinjuku NS building. In modelling many buildings, Nikken Seikei, Japan's largest design and engineering consultancy, found that the appropriate use of intelligent building technologies could lead to energy savings of 30 per cent.

4.5. The 1990s—building lifecycles, obsolescence and re-use

4.5.1. Recession and recovery

In the early 1990s the property markets moved into a deep recession in Britain and many other parts of Europe, Japan and the USA. The end of the office development boom resulted in an enormous oversupply of office space in many cities around the world. The phenomenon was widespread in Europe, from London, Paris and Amsterdam to Stockholm and also affected the USA, particularly New York, and Australia, especially Sydney and Melbourne. Japan was also affected where the so-called 'bubble economy', based on high property values, burst.

There were several interrelated aspects of the slump, including oversupply of offices and high vacancy rates, falling returns from property investment and rising property company bankruptcies, together with high exposure of banks to property loans. In London, by mid 1992 over 1 million square metres (18 per cent) of office space were vacant, compared to only about 370 000 square metres in the previous property slump in the 1970s. Property companies had borrowed heavily from banks and financial institutions and now owed debts of more than £40 billion. These circumstances temporarily stifled the markets for the construction of new types of buildings for the digital age. In Europe, Britain's position of pre-eminence in office development was undermined as property markets collapsed further and faster than elsewhere.

The collapse in demand for office space and slump in general property markets coincided with overall economic recession and this exacerbated the property market slump for three reasons. First, the demand for space diminished as firms went out of business. Second, many occupiers required less space as they trimmed their operations in order to survive. Third, firms in older properties were reluctant to move to newer, higher grade premises at a time when they were seeking to reduce expenditure. The costs involved in moving could be prohibitive and firms were often unable to change their existing lease agreements.

The property development business itself faced a new and uncertain future at the beginning of the 1990s. In Britain, developers turned their attention to new markets such as refurbishing older buildings and constructing smaller properties for niche markets. Those trader developers who had survived the threat of bankruptcy attempted to differentiate their buildings and services in a highly competitive market. They showed a growing awareness of the different needs of occupiers, including their requirements for information and communications technologies. They shifted away from the idea of purely speculative development, preferring to seek collaborative ventures with future tenants in the hope of ensuring that buildings would satisfy user requirements, thus providing a steady revenue stream.

The amount of available office space peaked in London in the summer of 1992 at about 3.2 million square metres. This represented a vacancy rate of 20 per cent. By mid 1993, London's office vacancy rates had fallen to nearly 18 per cent, including 38 per cent of offices vacant in the Docklands area (*Financial Times*, 13 August 1993, 16 December 1994). Vacancy rates continued to fall thereafter. By 1994, the beginning of an upturn in property markets was discernible. Private investors, especially from outside Britain, returned to the market, prompted by a fall in the value of the pound. Economic recovery and the movement of tenants from lower grade to higher grade premises meant that the surplus of top-quality offices began to fall substantially by the mid 1990s.

4.5.2. Obsolescence, adaptation and re-use

In Britain, the 1980s witnessed the highest levels of trading in buildings ever experienced. Technical obsolescence due to the inability of old buildings to accommodate modern information and knowledge-based work meant that many properties were no longer prime investments. Investors and owners disposed of these. A large stock of second and third grade buildings became available for refurbishment or demolition as financial institutions restructured their portfolios towards more modern buildings. The rate of obsolescence in office buildings increased as lifecycles declined from between 40 and 50 years in the 1950s and 1960s, to between 20 and 25 years in the 1980s. Since then they have continued to fall, boosting the potential stock of redundant office buildings. It was estimated that in December 1991, 220 000 square metres of office space in the City of London were virtually unusable and therefore unlettable. This figure rose to some 500 000 square metres by 1995 (Gann and Barlow 1996).

The boom and bust cycle, together with oversupply of office space and demand for buildings to accommodate information and communication technologies, had increased the rate of obsolescence of machine age buildings. Tenants had greater choice over the type of properties they occupied, generally preferring the more modern higher grade stock constructed in the 1980s. They also demanded changes to the structure of leases, such as the introduction of shorter term, more flexible leasing arrangements and the removal of upward-only rent review clauses. This helped to stimulate a trend away from the let-and-forget approach to property leasing towards 'cradle-to-grave' facilities and asset management, in which new services were offered to enable tenants to gain more utility from their buildings. Nevertheless, many users were inefficient in their space utilization and there was much room for improvement.

New working patterns meant that buildings had to accommodate occupiers' organizational changes, with the result that lifecycles of different parts of buildings were becoming shorter. The length of service of internal fixtures and fittings reduced so that in many buildings mechanical and electrical systems would be replaced every 10 to 15 years; partitioning, flooring and ceilings every five years; and workstations and cabling continually being moved. The idea of accommodating different levels of permanence and change in the built environment, characterized by different lifecycles of structures, fixtures and fittings, became a reality in the design and construction of shell and core offices during the 1980s.

But what could be done with obsolete warehouses, banks, offices and other machine age buildings? In central London alone, empty office space amounted to the equivalent of 35 Canary Wharf towers (1992). At the same time there was a need to create thousands of new homes, including 100 000 affordable housing units a year. This combination of trends in commercial office supply and use, together with the need to restructure housing provision formed the environment within which there was a potential to convert offices into flats. But it was difficult to realize this potential. Political, institutional and regulatory mechanisms hampered the change of use from one purpose to another. Many property owners and town planners failed to recognize the extent to which the use of space and technology was changing in the digital age. Owners often preferred to keep redundant buildings unoccupied in the hope that the market would change and planners often wanted to protect space for local employment and were therefore reluctant to grant permission for changes in use. In other cases, buildings were technically difficult to convert and the option was not economically viable. As Stewart Brand notes (1994, p. 2):

> *almost no buildings adapt well. They're designed not to adapt, budgeted and financed not to, constructed not to, administered not to, maintained not to, regulated and taxed not to, even remodelled not to. But all buildings (except monuments) adapt anyway, however poorly, because usage in and around them is changing constantly. When we deal with buildings we deal with decisions taken long ago for remote reasons.*

In spite of these difficulties, between 1993 and 1998, several hundred obsolete buildings were successfully converted in London and other major British cities, supplying around 10 000 new housing units. Figure 4.9 shows the Beaux Arts Building in Islington, London. This building was initially the Post Office Bank headquarters and was later used by British Telecom as offices, before being converted into luxury apartments. The number of conversions were, however, too small to make a significant impact on meeting the demand for new housing (Barlow and Gann 1993, 1995). By the late 1990s, the number of conversions to residential use in central London was already beginning to slow as developers turned to other, more profitable uses, including shops and hotels (Barlow *et al.* 1998). Location often made obsolete buildings attractive for conversion to other purpose, perhaps the best example being the change of use of high street banks into wine bars and cafes. In this case, building use is brought full circle, because banks were first established by people meeting in clubs and cafes to discuss their financial affairs.

Fig. 4.9. Beaux Arts Building, London

In Europe, and particularly in Britain, society attributes a sense of security and permanence to the buildings and structures that make up the fabric of our cities. Buildings represent the particular ways in which we use space and time. In the digital age, time and space are both being compressed. The built environment is continuously being changed to accommodate emerging social, economic and technical developments associated with these changes. Around half of all construction activity in most OECD countries involves work to existing buildings and structures, adapting, repairing and maintaining the built environment. The legacy of buildings produced during previous periods of economic growth, at different stages of the machine age, means that there is often a need to demolish old buildings and structures, or to make major modifications to them to accommodate the needs of the digital age. These processes of change illustrate acts of creative destruction brought about through the forms of innovation discussed in Chapter 1. Traditions and cultures, regulations and planning laws often retard the rate of change, whilst new economic and technical imperatives stimulate the need to innovate.

4.6. Summary

The machine and digital ages are similar in terms of the ways in which new technology and economic growth have created the need for completely new types of buildings, increasing the rates of obsolescence in the built environment. One technical attribute of the machine age was the discovery, production and widespread use of new materials such as iron and steel. These materials were used in the construction of buildings, factories and infrastructures needed by the rapidly growing industrial producers of such materials themselves. The digital age brings with it a similar phenomenon. The early development of intelligent buildings was characterized by the introduction of digital systems in buildings needed by telecommunications, computing and electronics companies, who were themselves the providers of the new systems which formed the backbone of the digital age.

During the 1980s, the locus of economic growth in OECD countries was shifting from that based on manufacturing towards the service sector in which information processing was becoming more important. Attempts to improve competitiveness and productivity of office workers through the use of information technology had effects as dramatic as those experienced in the automation of factory work in the first machine age. Indeed, companies ranging from IBM in the USA, Glaxo-Wellcome in the UK, SAS in Sweden and Sumitomo Trust Bank in Japan aimed to utilize new forms of buildings and associated technologies to achieve dramatic improvements in service sector productivity. Large multinationals began to consider their office buildings in much the same way that manufacturing firms considered their machinery and factories. Offices represented fixed capital investment in which the utilization of space, equipment and furniture should be maximized. The drive to improve efficiency in use stimulated major changes in design and layout. For example, in the early 1990s, IBM began to implement its 'hot-desk' policy whereby new buildings were to be constructed with 35 per cent less space than would have traditionally been thought necessary. This decision was made after surveys showed that sales and marketing staff only occupied their desks for 65 per cent of the working day. Instead of having a desk or office each, employees now had to find a vacant position and plug-in to an electronic information system to carry out their daily work.

A period of rapid innovation in the types of buildings needed for the digital age ensued. The most important manifestation of these changes was the development of new flexible, multi-functional offices and control centres. Retail developments, including out-of-town shopping complexes, the construction of leisure facilities and travel hubs such as airports also represented changing social and economic patterns. Many existing buildings and structures failed to meet users' new requirements. Buildings constructed in the 1950s, 1960s and early 1970s became obsolete because it was not possible to install new information and communication technologies and associated mechanical and electrical systems.

The period was also characterized by the rise of powerful private sector clients and sophisticated users who began to assume a more dominant role in dictating the pace of change in large projects. The development of much of the infrastructure towards the end of the machine age had been within the sphere of the public sector. Privatization of utilities and services, together with new forms of private finance, growth of international markets and demand for new production facilities meant that the private sector took a leading role in forcing the pace of change. In London, for example, developers such as Stanhope on the Broadgate project, and Olympia and York at Canary Wharf, were not prepared to accept the construction practices adopted during the previous era of publicly funded projects. They expected a higher level of performance, including contracts to be completed on time, within budget and to an acceptable level of quality. Some of these developers had considerable knowledge about construction technologies and processes and were prepared to intervene at a detailed level by appointing project management teams to advise on cost-cutting and time-saving methods. They set the British construction sector an ultimatum: if it was unable to meet performance requirements it had to change and if it proved to be incapable of change then firms from abroad would be used. The result was a steep rise in imports of materials and components, further internationalizing construction activity. The ways in which these pressures affect the processes of making buildings is the subject of Chapter 6, but first, specific processes of innovation in buildings for the digital age are explored in Chapter 5.

5. Complex constructs

This Chapter focuses on technological change in products, components, networks and systems during two phases in the evolution of intelligent buildings in Europe, Japan and the USA. First generation intelligent buildings were developed during the 1980s and they included electronic equipment to monitor and control internal environments and digital communication systems linking users to new local and long-distance networks. This phase was largely driven by technology suppliers seeking new markets for their systems, but there was little knowledge about how to integrate information and communication systems to meet user needs, or to create adaptive, responsive environments. The experience of using these systems led to questions about which technologies were appropriate for different types of users and for serving the needs of a diverse range of activities.

By the 1990s, a more sophisticated approach to the installation and use of electronic systems in buildings was emerging. The design and construction of second generation intelligent buildings placed greater emphasis on integration, adaptability and the management of space. This provided a more user-oriented, holistic approach to the installation of information systems, stemming from a recognition by designers and building occupants of the relationship between the use of space and new patterns of work. If users were to exploit the potential of information and communication technologies to the full, the development of systems needed to be related more closely to the configuration and use of space. As a consequence, a new trend emerged towards closer collaboration between users and producers in the design, operation and adaptation of buildings and the technical systems within them.

This Chapter begins by describing product complexity in relation to buildings and structures. It continues with an analysis of the rise of first generation intelligent buildings, focusing on building information systems, user systems and requirements for systems integration. It ends with a discussion of emergent second generation intelligent buildings and speculates how these might be designed in future. Chapter 6 explores design and integration processes, showing how complex products are put together, analysing the development of new process technologies and organizational changes in construction. Whilst the main emphasis in this part of the book is on the digital age with the introduction of information and communication technologies into buildings and structures, many changes have also occurred in materials, mechanical and electrical systems, structures and facades of buildings. It is not possible to review all of these developments here, but a number of technical changes are discussed where they illustrate how innovation in one field relates to changes in another, adding to the complexity of constructed products.

5.1. Product complexity

More demanding requirements of users and the development of sophisticated construction technologies have resulted in buildings and structures becoming more complex. Product complexity has a number of characteristics and dimensions (for a general discussion see works by Hobday (1998) and Hobday and Rush (1999)). For example, from a user perspective, it relates to the accommodation of multiple and often concurrent activities which alter as occupant requirements change over time. Many buildings and structures produced during the machine age exhibited a range of characteristics of complexity. These features appear to have multiplied in the digital age where problems in design, engineering and production have increased because of the need to take account of a growing range of variables relating to how people use space over time.

There is a difference between complexity of purpose and complexity in the means to achieving that purpose. Venturi argues that many building projects are complex in scope (such as hospital buildings or research laboratories) but even the house, whilst simple in scope, is complex in use, particularly if the ambiguities of contemporary experience are expressed (Venturi 1977, p. 19). Housing in the digital age is also becoming more complex in scope, particularly when there are requirements to accommodate digital infrastructures to facilitate new and varied functions (Gann *et al.* 1999). Most complex products have high initial unit cost and they are usually associated with long lifecycles, with requirements for periodic adaptation and modernization. But in many cases, lifecycles appear to becoming shorter. Some constructed products, like silicon chip fabrication factories, described in Chapter 4, represent large investments and have short lifespans.

From a producer perspective, the environment within which buildings are designed and constructed may lead to complexity because of the need to comply with a variety of regulations and because of the ways in which processes are coordinated. There are also a number of technical dimensions to complexity, including the range and types of technologies brought together in sub-systems and systems. Complex products usually have multiple and complicated component interfaces and are made of many tailored parts and sub-systems, which often interact in a non-simple way (for an early discussion see Simon (1962)). This means that for their production, complex buildings usually require many different specialist knowledge inputs, adding complexity in coordinating processes of systems integration. When new technologies emerge, this may give rise to the need for new specialist and integration skills, as discussed in Chapter 6.

In the digital age, complex products rely upon information, communication and control systems for their operation and this means that embedded software has become a core technology. These products and systems tend to exhibit non-linear properties through time in that small changes in one part of a system's design can often lead to larger changes in other parts. This is particularly the case when buildings are adapted to accommodate future uses or where further changes are made in response to user feedback. Digital technologies are being overlayed and intertwined within existing products, networks and systems. From a technical point of view, complexity is introduced because passive technologies making up the structure and fabric of buildings increasingly have to accommodate active technologies, including pneumatic, mechanical, electrical and electronic systems. The diversity and sophistication of components in such systems in buildings is similar to those found in ships or aircraft.

Moreover, in the digital age, the distinction between passive and active parts of buildings is becoming blurred. Modern buildings behave as dynamic systems, responding sometimes quickly and often too slowly, to circumstances created by changing patterns of use and physical conditions associated with their external climate (Groák 1992, p. 19). 'Smart skins', capable of changing their properties in response to the needs of saving energy in different climatic conditions, are being developed for building facades. Interactive facades can also be used to change internal and external appearances of buildings; these are constructed from new types of high emissivity and electro-voltaic glass. Piezoelectric materials can be used in building skins to detect pressure waves and help in acoustic tuning, using 'intelligent materials' to enable skins to revert to shape after movement (Noble 1997). Lightweight structures are increasingly being used. In some cases these are like tents, made from membranes of coated textile materials, they may be air-supported, pneumatic, or cable-net structures (Forster 1997). The Millennium Dome in London is a well known example.

'Smart structures', including mechanical elements and electronic controls enabling them to respond to movement such as wind loads or earthquakes, are being devised. For example, the Osaka Glass Tower (Fig. 5.1) has pendulum-like tanks of water stored at the top of the building for fire fighting. These are also used to damp oscillations in building movement caused by wind pressure or earthquakes. The Tokyo Gas Building (Fig. 5.1) has

Fig. 5.1. Tokyo Gas Building and Osaka Glass Tower, Japan

a different arrangement of digitally controlled counterweights in the roof. Old buildings are also being fitted with modern earthquake damping technologies, creating complex engineering and construction tasks. Figure 5.2 shows base-plate isolation techniques being deployed under the San Francisco City Hall. Designs for buildings with inherently unstable architectures are at the forefront of research into new structures and fabrics for buildings. The aim is to provide advantages in running costs and flexibility in use to accommodate many different functions. Research on these technologies, systems and structures parallels that on the development of unstable aircraft and other complex products—it would not be possible without the use of sophisticated and powerful computer simulation and modelling tools, themselves products of the digital age.

Skilful integration of new technologies by designers and engineers creates possibilities for the production of more flexible buildings. But the pace of technological and spatial change has not been evenly spread across markets for different types of buildings and structures and the spectrum of technical complexity therefore varies by product type. Larger buildings

Fig. 5.2. Base-plate isolation technologies fitted under San Francisco City Hall (1996)

tend to utilize a greater range of technologies than smaller buildings. For example, tall buildings and those with a depth of more than 15 metres are more likely to require air conditioning, lifts and elevators, sophisticated lighting, power and communications systems, and fire and security systems.

Tall buildings represent a particular form of complex product. They are required in part to meet the growing pressures on land utilization in areas of rapid urbanization where population densities are high. They are also symbols of wealth and status. From a cultural point of view, tall buildings may also represent a sense of 'the sublime' (Nye 1994). Perhaps for these reasons many of the world's tallest buildings are being constructed in Asia, whilst in Europe there appears to be a preference for smaller scale development. Nevertheless, the ability to engineer extreme slenderness ratios in tall buildings has improved following the discovery of techniques to damp vibrations caused by wind or ground movement. The use of strong, lightweight materials such as new forms of concrete (the strength of which doubled between 1975 and 1995) and high-powered computational analysis make it possible to produce buildings that are much taller than those of the machine age (Fitzpatrick 1997).

From a technical viewpoint, tall buildings are complex products because they require sophisticated electromechanical and control systems in order to manage internal environments, linked with rapid vertical transportation systems to move people and goods around them. Recent developments in elevators and lifts include the ability to combine vertical and horizontal transport within buildings, coupled with rapid response times using neural network computer control systems (Edwards 1998).

5.2. First generation intelligent buildings

From the 1970s onwards there was rapid technological innovation in the mechanical and electrical systems installed within buildings. Two features characterized these changes: an increase in the fusion of mechanical and electrical sub-systems, forming electromechanical systems, and the addition of microelectronic control technologies. The rising cost of mechanical and electrical installations as a proportion of total construction costs indicates their growing importance. In Britain, for example, the value of mechanical and electrical installations as a proportion of all construction work increased from 13 per cent in 1973 to 20 per cent in 1986 and has remained a little over 20 per cent since then (DETR various dates). Tables 5.1, 5.2 and 5.3 show estimates of the typical costs of mechanical and electrical systems as proportions of the value of total new construction, the value of refitting systems into existing buildings, and as a proportion of running costs in different types of buildings. Cost variations within different categories of building may be quite

*Table 5.1. Total value of mechanical and electrical systems as a proportion of total construction costs (sources: Harris (1988); *estimates based on interviews)*

Type of building	Proportion of cost: %
Air-conditioned office:	
375 000 sq. ft. (excl. underfloor power)	20
31 000 sq. ft.	18
Warehouse	
127 000 sq. ft.	20
Retail development	
276 000 sq. ft.	16
Luxury apartments	
65 000 sq. ft.	14
Highly serviced 'intelligent building' space*	40–65
Hospital*	30–65
Public buildings*	20–40
Shopping mall (basic core and shell)*	10–15
Housing*	5–10

pronounced, depending upon the type of building use and degree of sophistication of equipment. The highest cost areas are found

Table 5.2. Total value of mechanical and electrical systems as a proportion of retrofit or fitting-out costs (source: Harris (1988))

Type of building	Proportion of cost: %
Air-conditioned office	
175 000 sq. ft.	52
40 000 sq. ft.	62

Table 5.3. Value of various mechanical and electrical systems as a proportion of total running costs of highly serviced offices (source: Financial Times, 14 September 1987)

System	Proportion of cost: %
Services	35.0
Energy	23.3
General maintenance	23.0
Building management	22.4
Security	12.5
Other	18.8

in complex buildings such as hospitals, clean air rooms, research laboratories, television and communications buildings, control centres and specialist offices such as dealer rooms.

The direct cost of mechanical and electrical systems only presents a partial indication of the changes taking place within buildings and the effects this has on construction and running costs. For example, the use of air conditioning systems usually results in higher costs in building erection, incurred from the allocation of additional floor space for plant rooms and accommodating ductwork in special ceiling voids. In addition, roof-mounted plant may necessitate a stronger structure to carry the extra loads. In one 10 000 square metre office building, the central plant required for variable-air-volume (VAV) air-conditioning equipment needed approximately 800 square metres. Constructing this space in the basement added £500 000 (1988 prices) to construction costs (Harris 1988). When measured over a 50-year lifetime of a building, such systems may be replaced between three and five times, adding considerably to whole life costs.

In the early 1980s, the development of intelligent buildings was characterized by attempts of equipment and systems suppliers to expand their markets and sell new technologies. This resulted in intense competition between manufacturers who developed new marketing initiatives. The 'intelligent building' concept began as a marketing ploy by ShareTech, the ill-fated joint venture between AT&T and United Technologies, established in the wake of the break up of the Bell telephone system in the early 1980s. Engineers and sales staff believed that new electronic technologies could reduce the cost of controlling many of the existing functions within buildings and at the same time, facilitate activities which would not have been possible before, for example, remote control and monitoring of electromechanical systems, the provision for greater flows of information between building users, and the opportunity to sell new value-added services associated with building management, over a single network. New market opportunities emerged into which firms from different sectors swarmed. The structure, organization and activities of the industry supplying electromechanical and controls systems changed rapidly, becoming more international. It was very different from the traditional electrical and mechanical industries of the machine age.

The intelligent building concept spread quickly from the USA throughout Europe and Japan in the mid 1980s, although the pattern of demand for such buildings varied between regions. This resulted in some important differences in the way in which technologies were developed and used. For example, markets in the USA were vendor-driven. Suppliers of telecommunication systems perceived an opportunity to expand their markets by providing value-added, Shared Tenant Services for firms leasing space in speculatively developed office buildings. Equipment manufacturers pushed systems into the market and established an Intelligent Building Institute (IBI) to promote the concept.

By contrast, developments in Europe took a different form. In Britain the supply side, whilst playing a dominant role in pushing technologies, had a different complexion. The construction of systems was often mediated through the use of consultants, sometimes known as systems integrators, who could choose from different proprietary systems and were more closely in touch with user needs (although this did not necessarily mean that they built better, more user-friendly systems). They established their own Intelligent Building Group which, unlike the IBI in the USA, did not necessarily reflect the technologies being pushed by proprietary equipment manufacturers. Britain was in the vanguard during the 1980s, in terms of developing more sophisticated buildings than in most of the rest of Europe and in establishing an industry association to promote the intelligent building concept. By 1992, the British Intelligent Building Group had expanded to become the European Intelligent Building Group.

In Japan the situation was different again. Here, many buildings were constructed for owner–occupiers, rather than speculatively by developers. It was therefore easier to determine future user requirements. Rich Japanese companies invested heavily in electronic technologies for their new buildings. Moreover, government played a bigger role than elsewhere in sponsoring the concept of intelligent buildings. Japanese construction firms vied with electronics, telecommunications and office equipment manufacturers to produce prestigious buildings. Technological competition was intense but at the same time firms had closer links with users, giving a better balance between technology-push and demand-pull (Gann 1992).

Markets in Japan and the USA differed considerably from those in Europe in that much more of the construction work was new-build, permitting completely new designs to accommodate digital technologies. In Europe, apart from the City of London, there was a tendency to adapt, retrofit and refurbish the large stock of existing buildings. In the USA and Japan, applications remained large in scale, and in Japan particularly, intelligent buildings were controlled from central computer stations. By contrast, during the 1980s, the British approach evolved into one where semi-autonomous control stations were distributed floor-by-floor throughout new buildings. This entailed the design and installation of smaller scale systems which operated on networks, offering individual users a greater variety of choice in control over their immediate space. The British approach provided a more flexible solution which could be added to and adapted more easily than many found in the USA or Japan.

Different approaches to the design and installation of intelligent building systems illustrate a transition away from the rigid, standardized approaches to building layout and facilities of the machine age. Such differences are analogous to changes which were occurring on the shopfloor in manufacturing industries. In manufacturing during the 1980s, industrial restructuring occurred with a shift away from standardized, uniform volume-production lines. These were replaced either by a tightening of central control with the addition of flexibility in product ranges (similar to the American or Japanese approach to intelligent building systems) or by more flexible, dispersed production techniques, in which smaller production units were linked in networks of producers. These were comparable to the example of distributed control systems found in many European intelligent buildings.

By the mid 1980s, a pattern in the development of intelligent building technologies could be observed, grouped into two types: building information systems and user information systems. The latter was subdivided into those providing communications within buildings and those facilitating links between different buildings. Table 5.4 illustrates the main functions within each group. Each group is made up of sub-systems, many of which previously existed and were used separately prior to the advent of microelectronic controls. In the initial phases of development, digital building control

Table 5.4. Main functions of intelligent building technologies

Intelligent building technologies	Main functions
1. *Building information systems*	Energy management, temperature control, humidity control, fire protection, lighting management, maintenance management, security management, access control, space planning and management
2. *User information systems* (a) communications and data processing within buildings	Local area networks (LANs), electronic mail, data processing, word processing, management reporting and executive information systems, document image processing, and other internal communications such as audio/visual, CAD/CAM networks
(b) communications between buildings	Digital Private Automated Branch Machine, routing cost analysis where the landlord acts as public utility for the building, teleconferencing, valud-added data services (VADS)

applications were designed with similar configurations to those used prior to the advent of electronics. new control technologies were 'bolted on' to existing systems in much the same way that electric power was overlaid upon the old distribution systems associated with steam in the early twentieth century.

The market for electronic controls equipment expanded rapidly during the 1980s and demand for cabling, equipment and components rose fast. Firms from different industries swarmed into these expanding markets, and in the initial phases of development the type of systems design was often dependent upon the historical background of the main designer and supplier firms. For example, computer network and data processing firms tended to prioritize the cabling architecture. This was perceived to be the infrastructure at the heart of the system, to which controls devices and equipment could be connected. Telecommunications firms also prioritized cable networks together with digital switching systems. By contrast, the more traditional building controls' manufacturers placed electronic controls equipment at the heart of their systems, suggesting that these could be used to operate equipment connected to a range of different cabling networks.

A number of forces driving change within the information and computing industries spurred the development and use of new intelligent building systems. These technological developments included very large scale integration (VLSI), optical transmission and processing, information engineering, the integration of voice and data systems, radio transmission technologies, digital signal processing, and new person/machine interfaces. They resulted in a shift away from reliance upon the framework of old pre-electronic technologies, and a higher degree of systems integration using direct digital control. The novelty of these innovations was the way in which technologies originating from different industries were combined together to form new systems. But these markets and technologies were immature. There was a lack of clearly defined standards and complete off-the-shelf systems were unavailable. Failures in use of systems installed in the 1980s suggested that suppliers and designers did not fully understand user needs. Furthermore, the market place was immature in that there was a lack of user understanding of the potential benefits available from intelligent building technologies (DEGW/Technibank 1992, p. 1).

5.2.1. Building information systems

The first electronic building control systems were built on mainframe computers in the late 1960s. During the 1970s minicomputers were used, but market growth only became rapid with the advent of cheap microcomputers. It was estimated that by 1991, programmable electronic controls accounted for over 40 per cent of all types of building controls sold in the non-housing market (Proplan Search and Evaluation Services).

Figure 5.3 illustrates the groups of technologies typically found in building information systems. These evolved through a number of stages of integration during the 1980s, and by the early 1990s, the most advanced forms of control were available in integrated building management systems. Each sub-system is made up of electronic control devices — information processing, storage, retrieval and display; the media through which information is transmitted; equipment to be controlled; and sensing devices.

There were two benefits of using programmable digital controllers. First, they could control processes more closely to theoretical operational optima than traditional hard-wired controllers. Second, they facilitated the collection of data, permitting a degree of self-diagnosis (Senker 1986). Rapid

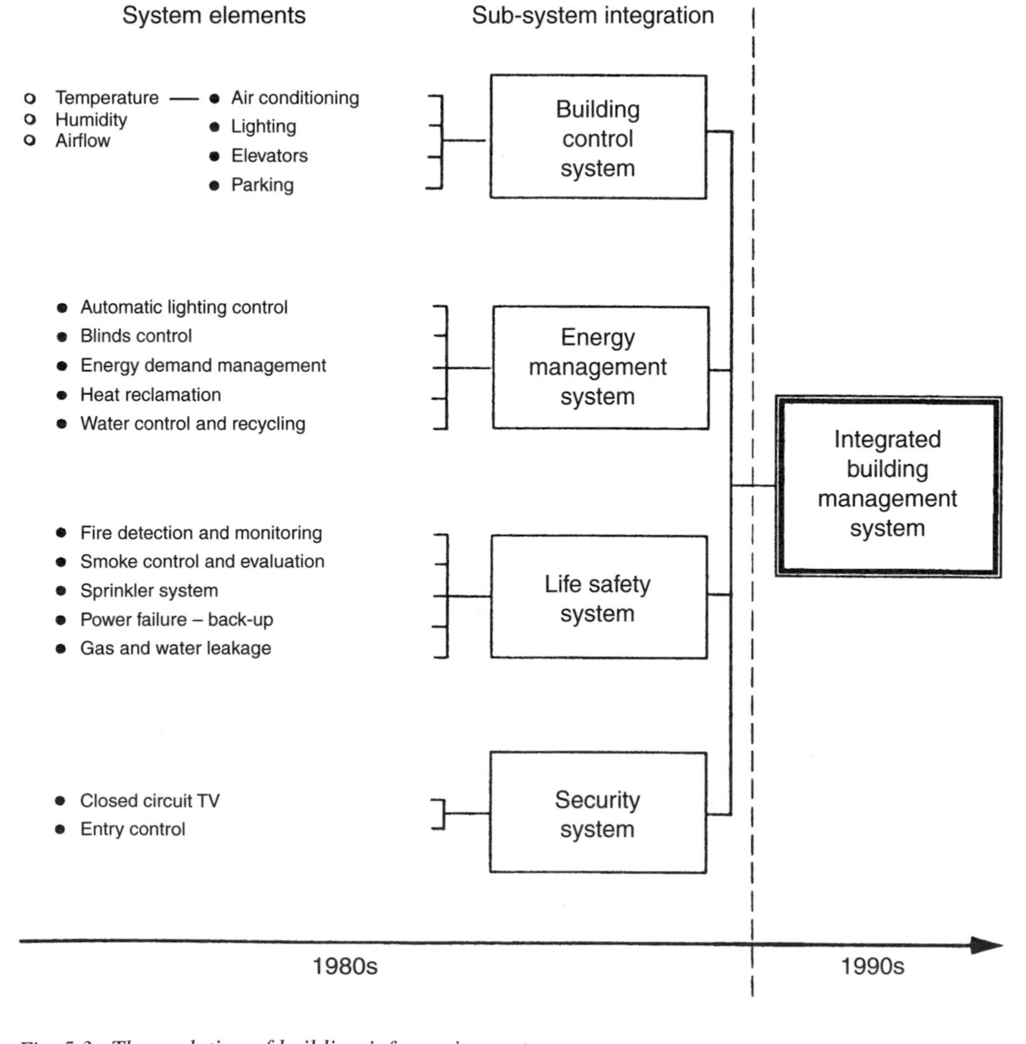

Fig. 5.3. The evolution of building information systems

market growth during the 1980s was associated with opportunities afforded by technological change, based on the development of direct digital control (DDC) and remote control and monitoring, which created greater flexibility in the application of control systems. Remote control meant that many buildings could be linked to one central computer, facilitating better energy monitoring by estates managers. For example, in the late 1980s, Berkshire County Council was one of the first to control energy used in over 200 schools from one energy management centre. Similar trends were found in chains of large retail stores, banks and pubs.

Substantial reductions in hardware costs and technical developments that increased the power and scope of equipment whilst reducing their size contributed to market expansion. Convergence between microelectronic controls and traditional modular controllers also increased the range of applications. A growing number of manufacturers of electromechanical equipment fitted electronic devices (such as programmable logic controllers) to their products to connect them to control networks. These replaced non-electronic components such as pneumatic sensors and actuators. Many types of equipment, such as air conditioning systems and pumps, were adapted to include microprocessors. The main benefits for users were reductions in running costs and greater flexibility in control. For example, 'smart pumps' permitted controlled flow rather than on/off switching; 'intelligent' lighting networks could reduce energy consumption, sensing whether rooms were occupied or not and turning lights on and off automatically. The total European market for multi-function direct intelligent control building information systems was estimated to be worth nearly £500 million by 1990 (Private Source).

Air conditioning became a standard feature in new office buildings constructed in the 1980s, replacing more traditional types of heating and ventilation. Variable-air-volume (VAV) air conditioning systems were widely used in Britain. These were an improvement on the air conditioning systems used in the 1960s and 1970s because they could respond more economically to local heat gains, although they had a limited range of controllability. They were cheaper to run because they provided savings on refrigeration energy and fan power. Prior to the introduction of VAV systems it was uneconomic to adjust the older constant-volume systems to locally changing conditions. Other systems, such as induction air conditioning, were often too noisy and took up too much perimeter space and ceiling fan-coil units gave poor air distribution and were associated with high maintenance costs (Bordass 1993, p. 86). Nevertheless, most air conditioning units installed in Europe in the 1980s relied upon large centralized plant and it was not until the early

1990s that engineers in Britain began to adopt an approach based on the use of small, distributed modular units which provided greater operational flexibility. This approach had been used for some time in Japan.

By the early 1990s it became more common to link energy management, security, access and disaster prevention into one integrated building management system (IBMS). The markets for these systems had expanded rapidly in Britain and parts of Europe during the late 1980s. New 'systems integrator' firms were established to design and install these systems. The technology facilitates the capability of centralized monitoring and greater control over the system as a whole. By 1991, IBMSs accounted for around 5 per cent of the British market for intelligent building controls and market share was growing swiftly. Integrated controllers usually operated by providing 'gateway' technologies linking various sub-systems, such as fire, access and security, lighting control and energy management, to central control units. Multiple tier systems architecture was used to provide what were claimed to be cheaper, more flexible and adaptable systems. More widespread use of open systems architectures helped to facilitate integration.

There was a strong element of supply-push in the evolution of markets for most types of building control systems. This was particularly the case in the USA where most developments were vendor-driven. However, building owners and users played a more prominent role in the demand for energy management systems. In Britain, energy use in buildings accounted for around half of total energy consumption and the energy crises of 1973/74 and 1979 resulted in large rises in energy prices. This, together with growing concern over energy conservation, stimulated the need for greater energy efficiency in buildings in the early 1980s. Markets for microelectronic energy management controllers grew rapidly during the 1980s, a growth of more than 300 per cent between 1981 and 1987, and a further doubling between 1987 and 1991 (Proplan 1985 and personal interviews).

It was estimated that 46 per cent of sales were in the office building market, 34 per cent in public buildings and 20 per cent in the industrial buildings sector, 30 per cent of the total market was located in the south east of England. About 75 per cent of sales were for fitting electronic controls to existing electromechanical systems. Technical changes facilitating the ability to retrofit controls and to add to the systems once they were in place, stimulated market growth. This trend occurred in spite of cheaper oil prices after 1986 which could have signalled a reduction in investment in energy management controls. Results of energy efficiency surveys showed that considerable energy savings were often achieved through the use of electronic energy management and lighting control systems (ETSU 1987). Such systems became smaller and cheaper. As a consequence, payback periods were shorter and it became cost effective to install them in smaller buildings.

There was a direct relationship between the increased use of information technology in office work and the demand for more sophisticated environmental control systems in buildings. The rapidly expanding array of information equipment produced more heat, resulting in the need for better air conditioning and energy management systems. In many large buildings the problem became one of how to manage energy used in cooling systems rather than for heating. The digital age brought with it two counterposing effects on buildings. On the one hand, new electronic office equipment was responsible for increasing heat gains which had to be removed; on the other, digital electronic control technologies provided part of the solution to managing these additional energy flows within buildings.

Buildings housing technically delicate and politically or commercially important information systems also needed additional protection against fire, theft and terrorism. A solution was again to be found in the use of electronic fire protection and security systems.

5.2.2. User information systems

User information systems installed within buildings provide the support infrastructure for internal communications and data processing in all types of buildings from offices, factories and hospitals to shops, hotels and leisure centres. In the machine age, information processing and exchange was largely a physical, face-to-face or paper-based process. Examples of mechanized paper transfer systems can be found in some large offices dating back to the early twentieth century, and the organization of typing work into pools had implications for space layout. But in general, in contrast to factories, office space was largely unconstrained by the needs of accommodating technology until the development and use of information and communication technology in the digital age.

Typical digital technologies developed for communications and data processing within buildings are listed in Table 5.5. Many of these were developed principally by office equipment manufacturers and electronics firms. Markets grew rapidly from the mid 1970s, spurred by the diffusion of photocopiers and fax machines, followed by desktop microcomputers. In Europe, the use of computers rose rapidly in offices from one terminal per office of 20 to 50 people in around 1975 to one shared by 4 or 5 people by about 1985. By 1990, many companies had provided one computer per person. There had been numerous developments in equipment for processing, distributing, storing and retrieving data. Integrated systems were intended to speed up and increase the volume of information flows, thus

Table 5.5. User information systems for communication within buildings
Network management
e-mail, intranet and in-house online services
Internet, intranet and extranet
Electronic room booking and conference facilities
e-commerce and in-house banking
CAD/CAM/CIM and simulation facilities
IT and software support
Online secretarial and administrative services
Databases and knowledge banks

attempting to raise productivity of office workers and meet requirements for increased information handling brought about by growth in the use of information technology. During the 1980s, computer equipment was increasingly being linked to Local Area Networks (LANs). As a consequence there was rapid growth in the volume of data cabling installed within buildings when compared with standard telecommunications and wiring for power.

Increased office automation gave rise to technical changes in cabling, such as the use of optical fibres for LANs and new cabling architectures such as structured wiring and new switching devices. These all required installation space within office buildings and the additional heat loads generated by higher densities of office equipment needed to be removed. In the 1970s, heat gains in a typical office building were about 20 W/m^2. By 1988, this figure had risen to 30–40 W/m^2. During the 1980s power requirements per computer rose to 300–400 W per unit. By the 1990s, the development of smaller laptop computers led to a reduction in power requirements to around 200 W per unit.

Despite the trend towards downsizing of equipment, the heat gains from office systems did not actually fall because reductions in power per unit were offset against an increase in the number of units installed. From about 1989, the trend towards downsizing of power requirements in office equipment meant that heat gains had reached a plateau. Nevertheless, during a ten-year period, the space taken up by cooling equipment had doubled. In large buildings the problem now was as much one of cooling in summer as it was the provision of heat during winter. Balancing heat gains and losses automatically required integrated systems. This created demand for new types of energy-saving buildings. For example, Sainsbury supermarkets were among the first to be designed to operate without any dedicated heat sources. Heating was derived from that given off by lighting and refrigeration equipment. Cooling and air conditioning became an essential complementary technology in the design of modern office buildings.

By the mid 1990s, cableless communications technologies were beginning to compete with those described above, providing savings on space required for installation of hard-wired systems and facilitating greater adaptability in the use of existing buildings.

The proliferation of new public and private networks contributed to another wave of change in the design, construction and use of buildings. Telecommunications is the third group of digital building technologies used for communications between buildings through advanced telecommunication ports which interconnect buildings with new broadband digital information highways. Figure 5.4 illustrates the main functional elements in these systems.

In the USA, the impetus toward the development of intelligent buildings was accelerated by deregulation of telecommunications in the early 1980s. This led to the emergence of companies offering shared tenant services (STS) as a new business to increase the value added from the investment in intelligent buildings. In Britain two events stimulated the growth of new digital networks: the liberalization of the telecommunications infrastructure in 1984 and deregulation of financial services (the Big Bang) in 1986. The liberalization of the telecommunications infrastructure was accompanied by growth in private digital networks, and a wider range of choice in public networks. In the late 1980s, Britain had more digital leased lines in use than any other European country, and there was further potential for a rapid expansion of value-added services following the Duopoly Review of 1990. Other European countries followed and by the mid 1990s there were many different service providers in operation.

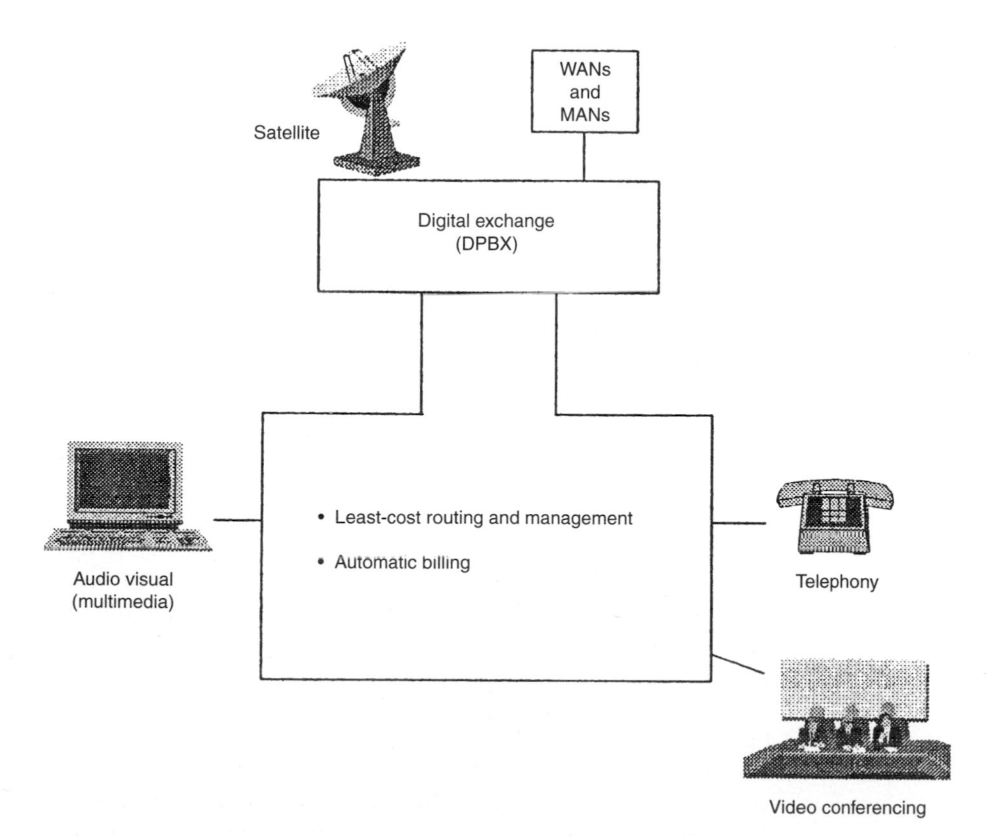

Fig. 5.4. User information systems for communication between buildings

Technological change in telecommunications linking internal networks with those outside was rapid. It included developments in switching and networking systems designed to link local area with wide area networks (WANs). Integrated Systems Digital Network (ISDN) and Intelligent Network (IN) technologies offered the potential for many novel applications, such as sophisticated voice and data switching. Programmable Digital Multiplexors (PDMXs), used to distribute intelligence out from the switch and into the network, were superseding the traditional public exchange. New information services were proliferating during the 1990s (e.g. video conferencing, video phones, remote access interactive video and third generation mobile phones) as US telecommunication firms began to compete in Britain and other European countries.

The installation of optical fibres in the long-distance telephone network continued in Europe, Japan and the USA through the 1980s and 1990s. Costs of optical fibres fell sufficiently to justify installing them in local loops (i.e. between the office or factory and the nearest telecommunication exchange) for high-revenue business customers. After 1990 it became common to install fibre optic risers into large new commercial office buildings. However, the rate of diffusion was slowed by the problem of converting from light to electric pulses cheaply: converters were initially relatively expensive.

5.2.3. Systems integration

Prior to the advent of microelectronics and information technology, building information and user information systems functioned in different spheres carried out on separate technical systems: information processing was predominantly a physical, face-to-face or paper-based process, and the internal environment was controlled mechanically. Since the mid 1980s, however, there has been a growing trend towards the integration of systems for economic reasons, facilitated by technical and organizational changes which provided cost savings to producers and/or to users.

Integration occurred at two levels. First, it occurred at a lower level, within each of the groups of technologies discussed above (for example, in the case of integrated building management systems, shown in Fig. 5.3). By the early 1990s this level of integration was well advanced in Britain. Second, at a higher level, the integration of elements between building automation, office automation and telecommunications was evident at least in terms of the ways in which systems functioned. This stage of integration is sometimes referred to as 'computer integrated building' (CIB) (DEGW/Technibank 1992, p. 5). In the early 1990s, Japanese developments were at the forefront.

The advent of CIB was made possible in part because of the diffusion of microelectronics into electromechanical subsystems, user appliances and telecommunications equipment.

This in theory provided a common technological medium, in the form of digital electronics, which could permit the flow of information between systems. At the same time there had been a growing functional interrelationship in the way that these systems were operated. One consequence of systems integration was that firms from the traditional mechanical and electrical sector now had to compete with firms from computing and telecommunications to provide both the enhanced communications infrastructures and more sophisticated environmental control.

Pressures to recoup sunk costs invested in intelligent building technologies gave rise to a new set of cost-reducing economies associated with the benefits of using integrated systems. Expenditure on plant and equipment in new and refurbished buildings increased the ratio of fixed over variable costs, such as wages and materials which vary with output. Fixed costs had to be paid even if the building system was not operating, or was only operating at low levels of capacity utilization. The rate of obsolescence of some intelligent building technologies was rapid and the lifecycle of electromechanical equipment such as heating and ventilating systems was generally accepted as around 15 years. But changing user needs and/or inadequate specification of capacity together with rapid technological change led to quick obsolescence. This meant that many systems had to be replaced, in some cases after a very short period of a few months. Building owners and occupiers were therefore reluctant to invest more than was absolutely necessary for integrated systems. But this in turn acted as a stimulus for technical change.

Size mattered in the cost/benefit equation. Initially, it was only cost effective to install integrated control systems in large buildings (over 1000 m^2). Economies of scale resulted from the number of users with access to the system. In remotely controlled systems, the number of buildings linked to a single remote control centre could be expanded, thereby reducing the marginal costs of adding each additional building. This was the main reason for the growth in the number of remote control and third party bureau services.

However, the costs of integrated systems were not related simply to numbers of users. Costs could be reduced if improvements in system performance were made. For example, if systems could handle larger volumes of information then fixed costs per unit of information would decline as the volume of information flowing around the system increased. For this to happen, improvements in transmission and or switching capacity needed to be made. This stimulated further innovation in integrated systems design to improve utilization and information throughput rates. For example, the installation of optical fibre LANs in buildings enabled much greater volumes of information to be transmitted more quickly than before.

Two further economic conditions stimulated systems integration in buildings. First, savings were achieved by using the same building management system to control lighting, air conditioning and access control. Initially these systems were wired separately and users controlled them from different devices. Their integration and provision over one cabling network with control from one device resulted in lower costs to users. Economies of scope were therefore achieved by using the same equipment and networks to serve a number of different purposes. Second, economies of system were achieved through advantages gained in controlling intelligent building networks from a single point of control rather than from several (Davies 1994). The development of network management, gateway and control software was important in realizing these savings on integrated systems. Here the emphasis was on reductions in costs of control to systems managers, rather than in the provision of existing and additional services to users at lower costs.

In Britain, the potential savings offered by operating many sub-systems on one integrated system exerted strong pressures towards integration within building management. This resulted in energy management, air conditioning, lighting and access control being managed from one central station, shown in Fig. 5.3.

One example is the 'head-end' management system that provides software interfaces to a number of different proprietary sub-systems facilitating control from one visual display unit. One of the leading designers is the British firm CdC (Central data Control)—the company that designed and installed systems in the Blue Water Development, Thurrock Lakeside Shopping Centre and the Bromley Glades Shopping Centre during the late 1980s and 1990s. These systems permit a degree of sub-system autonomy in that different elements or different zones within a building are controlled by semi-autonomous outstations. These interface with the central control station but can continue to function in the event of a central failure. In Britain there is a trend towards taking this form of networking one stage further so that a building is controlled by a series of distributed substations each capable of providing some centralized control function if necessary, thereby reducing the risk of catastrophic systems failure. The use of electronic control systems, however, created its own operational dangers similar to those found in the application of electronics to aircraft and air traffic control: the problem of 'fly-by-wire', in which decisions are taken by computer in the event of disaster. Figure 5.5 illustrates different sub-systems in a fire detection and prevention system installed within a modern shopping complex or intelligent office building. Each of the gateways between sub-systems is linked using software interfaces. The danger here is that unpredictable outcomes might occur as a result of bugs in the software or because of the ways in which building users and managers respond to preprogrammed instructions.

Despite the theoretical economic advantages of integrating digital systems, there was little evidence of higher levels of integration between building management, office automation and telecommunication systems during the 1990s. Each was generally contained separately by three different types of facility

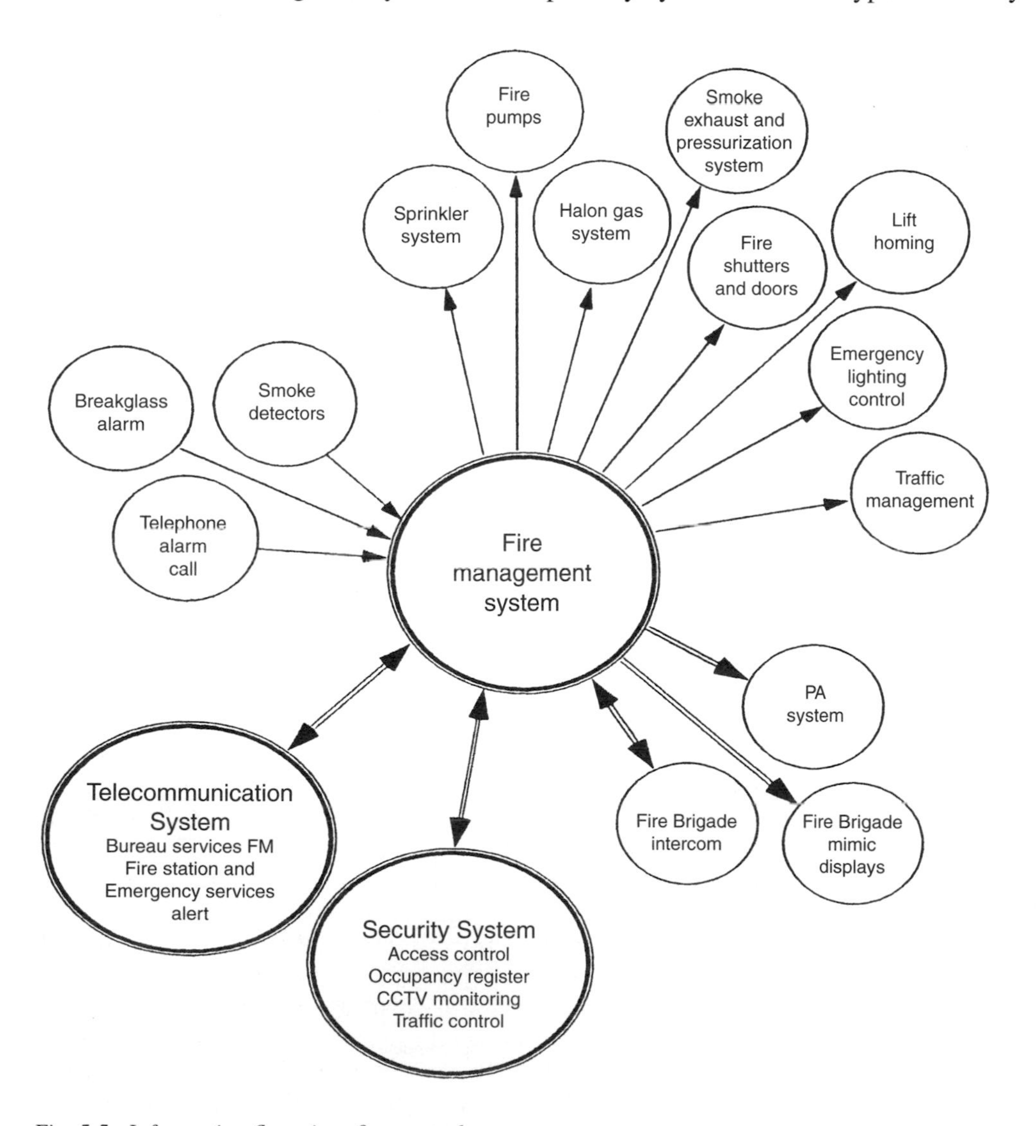

Fig. 5.5. Information flows in a fire control system

manager: building managers, data processing managers and telecommunications managers. Nevertheless, an increasing number of systems were designed with information gateways between building automation, office automation and telecommunications systems, such as those shown in the fire control example.

Japan appeared to present an anomaly in the 1990s, where high-level integration on large-scale centralized systems often controlled by mainframe computers could be found in some buildings (see Fig. 5.6). Perhaps this reflected different paths of development of computing and of building use. This type of integration was only achieved in large buildings in which there were thousands of users: the explicit aims being to reduce running costs, increase productivity of office workers and

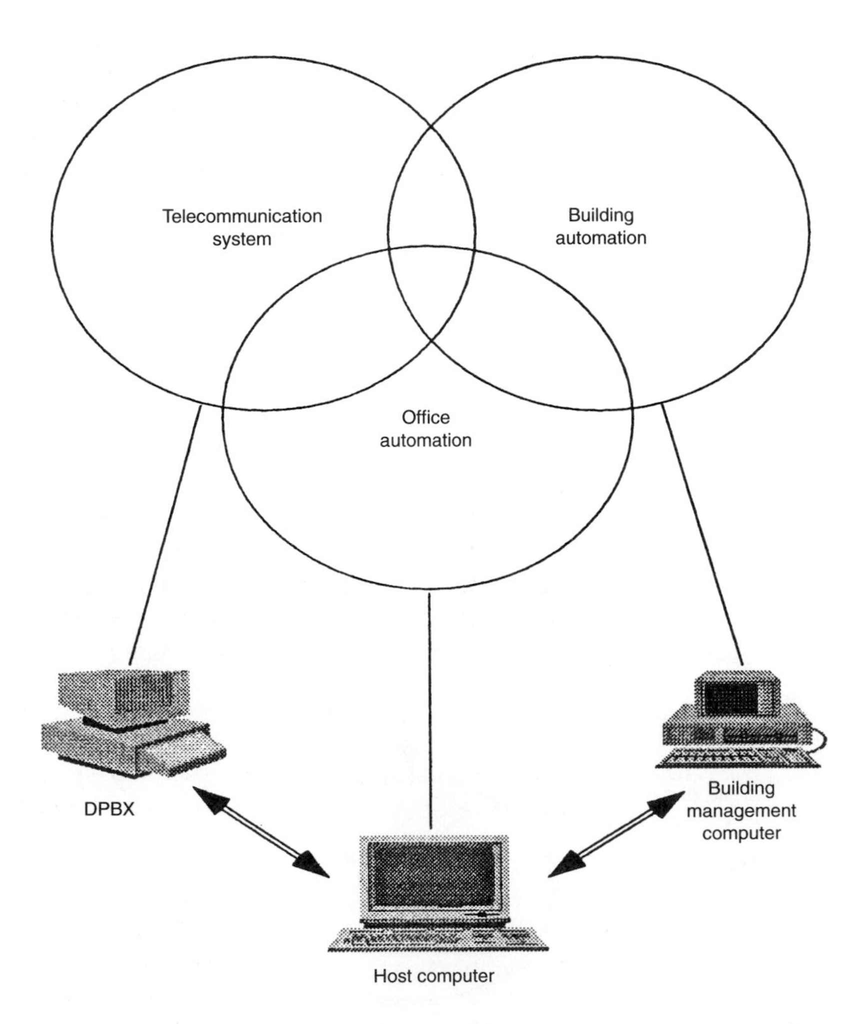

Fig. 5.6. Example of high-level systems integration in a computer integrated building in Japan

provide access to new information services over a single network. Such buildings were constructed at enormous costs to powerful Japanese corporations as showcases to demonstrate their technological capabilities. Integration was usually achieved technically by the provision of gateways between the building management computer, the telecommunications switching system (DPBX) and a host computer serving the office automation system.

The pace of systems integration was slower than it might have been for four reasons. First, because the abilities of systems designers were often stretched to overcome difficulties of introducing digital controls in sub-systems, let alone developing higher level integrated systems. Second, because there were bottlenecks in capacity utilization, and lack of organizational change to cope with new forms of operation. The most important problems of integrating building systems were often organizational rather than technical or economic.

Third, the institutional framework which evolved prior to systems integration preserved the methods of operating with traditional hard-wired technologies and was out of step with the requirements of digital building systems. Building regulations covering existing practices took time to revise, and changes in building control therefore tended to lag behind developments in new technology, slowing the rate at which new techniques can be adopted. Insurance requirements may have also retarded the adoption and diffusion of new technology. For example, in the 1980s, many insurance companies were reluctant to provide cover on buildings which relied upon electronic fire protection systems, stipulating that stand-alone, tried and tested hard-wired back-up systems must also be installed. This negated the potential economic and operational benefits which could be achieved from integrated electronic systems. The institutional environment within which intelligent building technologies developed was therefore complex. For example, some of the factors constraining innovation in one direction may also stimulate change in another: regulations concerning environmental protection have contributed to the need to install better energy management systems. Furthermore, insurance companies have in part stimulated the use of electronic security equipment by providing cheaper insurance cover for buildings with comprehensive monitoring systems.

Fourth, integration is hampered by the difficulty of providing compatibility and achieving interconnectivity between subsystems. Problems associated with systems compatibility and pressures to provide better integration between networks have given rise to the need for standard operating procedures and protocols and/or the development of gateway technologies. In multiple occupancy buildings, lack of common standards can result in incompatibility between equipment installed by the landlord and that which tenants wish to install. Different standards have evolved and setting or changing standards can be costly, arduous and time-consuming. Table 5.6 provides a summary of the major constraints to systems integration in Britain.

The failure of many first generation intelligent buildings to provide significant returns on investment suggested that suppliers and designers did not appreciate the nature of demand. Neither did they fully understand how to develop systems to

Table 5.6. Constraints to systems integration

Constraint	Type
Problems with hardware and software reliability	Technical
Slow technical development of distributed intelligent controllers and multifunction central stations	Technical
Lack of process innovations by the construction sector, which could facilitate better installation of new technologies	Technical and economic
Short product life but long lead times in product innovation requiring large, long-term investment in R&D	Economic
Users locked in to single systems requiring expensive bespoke protocol converters	Economic
Lack of hardware and software compatibility and common standards	Regulatory and technical
Out-of-date building and fire regulations devised for electromechanical rather than electronic controls	Regulatory/ institutional
Conservative approach taken by insurance companies towards the use of some new technologies where new risks may be involved, such as in electronic fire systems	Institutional
Fragmented structure and practices of the construction sector which hamper the flow of technical information about how to design and install integrated systems	Skills and organizational
Structure of end-user organizations based on mode of operating with single discrete systems and incompatible with integrated systems, causing institutional lag	Skills and organizational
Nature of property development and management, based on 'let-and-forget' speculative property development, in which financier, developer and letting agent have little knowledge or interest in end-users' needs	Organizational and institutional

meet users' needs. Evidence of this could be found in the literature on the design of first generation intelligent buildings, which focused on technical issues such as cabling layouts, the integration of hardware and software, switching systems and the requirements of access space for system installation. These issues were set in the context of how to fit new technologies into existing types of buildings. There were few attempts to develop a new architecture to meet the requirements of working with information technology (Kroner 1989). The implications of these technologies for users were hardly mentioned, although the rationale for investing in such technologies was to improve work place environments.

The market place was itself immature. Users did not generally understand the potential benefits which could be provided by specifying intelligent building technologies. Most users failed to relate investments in new technology to organizational changes and their use of space. Most suppliers were incapable of understanding these issues because they were mainly concerned with technical matters and had little knowledge about users' needs.

5.3. Second generation intelligent buildings

First generation intelligent buildings were defined primarily by their technical elements. The main approaches to their development in different countries are shown in Table 5.7. Many of the technologies initially introduced in buildings for sophisticated users in telecommunications and computing firms or the financial services sector were adapted for other user groups, but it was by no means clear that these systems were actually appropriate for other applications. The problem was that few users were experienced enough to articulate their needs and most had to rely upon the experience of equipment suppliers and information systems designers.

The term 'intelligent building' was misleading and poorly defined by practitioners and this often resulted in confusion and false expectations on the part of users. In many cases the term was used as a marketing ploy by letting agents wishing to maximize the rental of buildings fitted with simple electronic air conditioning controls or computer cabling systems. Many users suffered from sick building syndrome, with symptoms such as headaches and allergies caused by dust, lighting and materials, and other complaints caused by tightly controlled and highly serviced buildings. Users became increasingly unhappy about the new environment of first generation intelligent buildings. These complaints gave rise to a countertrend away from the construction of highly serviced buildings towards a reliance upon low-technology well designed buildings where passive systems such as the fabric, structure and layout could be used to provide energy efficiency and a comfortable environment.

Table 5.7. Supply and use of first generation intelligent building technologies in Europe, the USA and Japan

	Europe	USA	Japan
Technology-push versus market-pull	Supply-push but some user representation through consultants and systems integrators acting as intermediaries	Supply-push but little user representation — vendors dominate market	Mixture of supply-push and demand-pull, more user-oriented. Competition between construction, electronics and telecommunications firms creating dynamic environment leading to many innovations
Role of end-users	A few examples of business-driven development where users play a major role	A few examples of business-driven development where users play a major role	Markets oriented towards end-users in owner-occupied buildings: business-driven approach
Central versus distributed systems	Smaller systems — trend towards distributed network systems based on PCs	Large, centralized systems for tenants based on shared-tenant services in speculative offices	Large, centralized, bespoke systems for owner-occupiers. Based on mainframe computers
Level of systems integration	Little integration between building, office and telecommunication systems, but trend towards integration in building management systems	Some integration between building, office and telecommunication systems	Higher level of integration between building, office and telecommunication systems

Moreover, in Europe, concerns over energy use grew, together with the need to control greenhouse gas emissions and this meant that owners and designers increasingly sought to produce buildings with natural ventilation.

In the 1980s, many firms had made huge investments in both building and user information systems. These systems proved successful in some areas — such as helping to improve energy efficiency — but hoped-for business improvements in productivity and access to new markets were not always achieved. In general, the service sector appeared to have been unable to exploit the potential benefits of investing in new technologies, and investment outstripped payback. Moreover, expensive equipment quickly became obsolete.

Productivity gains in new applications such as fully networked offices were elusive, although some of the better new buildings provided more flexible spatial arrangements. In general, new technology had been used to interconnect different machines but the number of users remained the same and their work rates had not risen. Furthermore, the ratio of professionals to support staff changed little during the 1980s. This was the productivity paradox in the service sector (Freeman and Soete 1990, OECD 1991). Studies of office productivity showed that productivity growth rates were slow during this period. This was sometimes caused by office managers who automated single tasks, but failed to automate complete office processes. For example, when faster fax machines or printers were purchased, they made little difference if they were located in post rooms from where it took several hours to sort and deliver messages (Touche Ross 1991).

Investment in new buildings and systems shifted firms' cost structures from variable to fixed, reducing flexibility without providing productivity increases (Roach 1991). Only a few organizations restructured their operations to benefit from investments in new systems. These included some international banks and multinationals such as IBM, whose SMART (space morale and remote technology) initiative was claimed to have resulted in productivity gains of 3 per cent in 1985 and 20 per cent by 1988. These companies were, however, the exceptions and the once hoped-for paperless office was a long way from realization.

Conditions changed in the early 1990s, with enormous white collar job losses in banks, insurance companies, and the retail sector because of the combined effects of recession and restructuring due to increased competition following deregulation. Office markets throughout OECD countries were hit by a deep recession in the early 1990s. The trend towards electronic office work began to reduce space requirements, in some instances by as much as 30 per cent, such as in IBM's SMART programme for their new buildings at Bedfont Lakes, outside London. If productivity improvements were to be achieved there might be even fewer jobs in the service sector, requiring less building space. This was to have implications for the ways in which firms used the technologies already installed in first generation intelligent buildings, some of which were to prove inappropriate for meeting future needs. Furthermore, technological developments were continuing and pressures for environmental protection—including new legislation—were resulting in changes to the ways in which buildings were designed, constructed and used.

First generation intelligent building technologies on their own were not sufficient to provide gains in productivity. The high rates of turnover of occupants meant that new ways of providing flexibility in building use had to be found, together with better

Fig. 5.7. Commerzbank Headquarters, Frankfurt (1997)

ways of integrating and installing digital technologies. This entailed closer links between building designers, property managers and users. By the mid 1990s, the term 'intelligent building' began to take on a new meaning. It was defined more broadly to include the efficiency and effectiveness with which buildings could be used. Emphasis turned away from the purely technical towards social and economic aspects of use, including the management of space. It became apparent that organizational changes were required, staff needed to be retrained and space had to be allocated in new ways. There was a change in direction away from the supply-push of the 1980s, towards a more user-oriented approach to the development and use of new buildings. New landmark facilities were constructed, but these attempted to marry spatial layouts with the use of technology for knowledge workers in environment-friendly buildings. The new headquarters of Commerzbank in Frankfurt, completed in May 1997, attempted to demonstrate these principles, combining 'sky gardens' with natural ventilation systems and sophisticated digital building control and information systems. It is the tallest office building in Europe, constructed in a city during the period when the German Green Party was in power (Fig. 5.7).

A series of research reports led by DEGW, the British firm of architects and space planners, in collaboration with the Building Research Establishment, engineers Ove Arup and cost consultants Northcroft, provided the first serious analysis of the ways in which buildings were changing in the digital age (see, for example, DEGW/Technibank 1992, Duffy *et al.* 1993, Duffy 1997, DEGW *et al.* 1998, Harrison *et al.* 1998). This work began in the 1980s with a series of studies on organizations, buildings and information technology (ORBIT). The work was important for three reasons. First, it provided analyses of the emerging relationships between the use of space, technology and work processes. Detailed case studies of buildings and work processes were carried out in Europe, Latin America and east Asia, which showed different ways of organizing space and utilizing information systems. Second, it provided new conceptual frameworks for evaluating building performance in the digital age. Third, it illustrated a new area of interdisciplinary professional consulting work in advising building occupants about how they could improve internal work environments and the efficiency and effectiveness with which their buildings are used. DEGW's mission of integrating people, processes and places represents a new approach to the design of work places. By the mid 1990s, a series of occupancy surveys by Leaman and Bordass (1997) provided further evidence of the links between office productivity, spatial design and building technologies. A new body of knowledge about the design and use of buildings in the digital age was emerging.

In the late 1990s, office building markets began to grow again as multinational firms such as Ford, IBM and American Express required new highly serviced space as they reorganized their businesses. They changed their use of buildings to compete in new European and other international markets. They sought new efficient ways of conducting business through networking operations, developing 'euro-offices' that utilized centralized control of information systems. Other firms followed and large, modern, multi-tenanted buildings were needed in central business districts in European capitals.

Buildings such as the headquarters for SAS outside Stockholm, completed in 1988, paved the way for second generation intelligent buildings. It provided convivial, comfortable places to work, with space to interact with colleagues, designed with internal streets and cafes, meeting places and information points. It was designed with knowledge-intensive work in mind, conscious of the need to meet high environmental standards. In Britain, British Airways followed suit with its Waterside building, as did Boots the Chemist with its headquarters in Nottingham, both completed in the late 1990s. These buildings attempted to link a more flexible culture and processes of knowledge-based work with information and communications technologies in buildings which had minimal impact on the environment.

Many second generation intelligent buildings were smaller than those constructed in the 1980s. Small buildings represented a potential market for companies supplying information and communication technologies and buildings with floor areas of less than 500 square metres accounted for over 80 per cent of all office premises. From a technical point of view, systems with a low installation impact on existing buildings needed to be developed. These began to emerge in the early 1990s and included cableless technologies such as radio and infrared, miniaturized equipment and simplified systems. They were more suitable for the major market for intelligent building technologies in Europe, which were in the retrofit of systems into existing buildings. Many users wished to upgrade their existing premises by retrofitting electronic technologies and installing new cabling systems. This took two forms: new systems were installed in existing buildings after old systems had been removed, or additions were made to existing systems through the installation of new equipment, for example, the addition of microelectronic controls to existing heating and air conditioning systems. The rate of change in electronic office equipment was so rapid that even one-year-old buildings were sometimes refitted with new cabling and ventilating equipment to cope with changing user needs. The retrofit market was also fuelled by the need to eliminate health risks found in older systems, such as those associated with asbestos in old heating systems. Many public sector buildings, for example, were refurbished to specifications that included the installation of electronic control systems. The extent to which retrofitting took place was, however, limited by the space available for the installation of new equipment.

Better building facilities management and space utilization planning aimed to help improve performance of buildings. The facilities management profession emerged in the early 1980s with much promise of providing professional expertise about how buildings could be used efficiently and effectively. However, the profession appeared to be lost in technical detail during the 1980s and early 1990s. In Britain, it appeared not to be able to express the values appropriate to the new knowledge-based society. Most buildings failed to delight from the points of view of both their design and management.

In the early 1990s some user organizations were beginning to change their use of office space in quite radical ways, to enable them to respond to market changes and to compete more effectively. For these organizations their space needs were no longer firmly rooted at specific locations. The 'placeless office' or 'virtual office' was becoming a reality in which organizations' communication needs took primacy over building requirements. The use of smaller, more decentralized cableless technologies grew and some of these needs began to be accommodated. This reduced the installation impact of information technology on buildings compared with experiences of the 1980s. For example, cordless radio-controlled fire detection systems were already being used in the early 1990s. This trend, together with that of downsizing, meant that heat gains were lower and the need for air conditioning equipment and other environmental controls had also reduced. By the mid 1990s there was already a divide in the design and engineering community and amongst property developers and letting agents—between those favouring the 1980s' solutions of heavily serviced, fully air conditioned buildings and those promoting low-energy designs relying upon natural ventilation. The latter approach reflected a shift towards the use of passive systems in buildings designed to have a lower impact on the natural environment, both in their construction and use.

5.4. Summary

First generation intelligent buildings were something of a false dawn for the digital age. They often resulted in the addition of another layer of technology within and between buildings, thus adding to their complexity. They also often reinforced hierarchies and functional divisions in organizations because they inherited spatial arrangements from the machine age. The development and use of first generation intelligent buildings was more successful in some countries than in others. The supply side dominated the North American approach and many buildings failed to meet user requirements. Britain led the rest of Europe in the development and adoption of new technologies but failed to meet user needs as successfully as the user-oriented approach found in other North European countries. The most successful buildings tended to be those constructed for owner occupation where user requirements could be specified in advance. Buildings constructed by property speculators did not generally meet the same standards in terms of space and equipment installations. The Japanese experience was important—by the late 1980s there were more intelligent buildings constructed in Japan than elsewhere. Japan was the location of much new investment, innovation and experimentation with new types of systems during the 1980s. These buildings were a vast improvement on most existing offices which traditionally provided far worse accommodation for information-intensive work than those in Europe and North America. It was not difficult, therefore, to achieve some of the benefits and improvements in productivity claimed by suppliers in the West.

In general, suppliers of first generation systems had painted rosy pictures about the future of work using intelligent building technologies, including the possibilities of organizing work in new ways and the construction of new more flexible buildings that could accommodate change and provide high levels of control over internal environments. In contrast to attempts to industrialize the fabric and structure of buildings and construction processes, developments in building control systems were not pursued by architects or general construction firms. Indeed, generally both were initially suspicious of such technologies. A separate industry evolved to measure, monitor and control functions within buildings. This industry expanded beyond functions of control, to provide building information systems. Firms from a number of different sectors converged on expanding markets, each bringing a specific approach conditioned by previous experience in other market sectors—for example, temperature controls, and process and manufacturing plant control.

The initial use of centralized control illustrated the way in which new technology is often juxtaposed upon old. It was not until the full implementation of electronic equipment in all parts of the system that major cost savings were made and this also increased the flexibility of space utilization in buildings. Moreover, the success of systems innovations was dependent upon improvements in many sub-systems. Developments in one technology had a direct effect on developments in another. For example, heat gains from electric lighting and office equipment had to be removed through the use of more advanced air conditioning systems. This innovation process resulted in a transformation in supplier and user organizations which was in many respects similar to that experienced with the introduction of electric power at the beginning of the twentieth century. Both changes facilitated the design of new types of buildings.

Second generation intelligent buildings represented a new approach, linking people, processes and places with technologies in buildings. Design firms like DEGW were at the forefront in creating new concepts about how to use space. Constructing effective and efficient second generation buildings also placed new demands on engineers, contractors and systems suppliers. Traditional national construction firms faced more competition from two new sources: equipment supply industries and innovative international construction organizations. At the same time, new user–producer relationships emerged. Designers and builders had to re-think their approach in order to operate in these markets. This is the subject of Chapter 6.

6. The new production system

The digital age brought about new opportunities and threats to producers. Changes in demand were leading to the construction of more sophisticated buildings incorporating complex mechanical, electrical and electronic systems. By the 1990s, property development and construction industries were confronted by three requirements:

- to construct buildings facilitating the provision of enhanced information services;
- to create flexible spaces accommodating the needs of information-intensive activities;
- to reduce the environmental impact of buildings, in part by including more sophisticated environmental control technologies in them.

These created strong pressures to change construction techniques. The rising proportion of high-value engineered components and systems used to produce the structures, facades and internal installations in buildings meant that the nature of construction work changed. There was an increase in the number of specialist activities and some areas became technically more demanding. Work on construction sites shifted further away from mixing, cutting and shaping, towards alignment, assembly and fixing. There were also changes in work organization, with the demise of general contractors who had previously employed skilled craftsmen directly. Instead, specialists were hired by prime contractors who now supplied project management services. Specialists either employed their own workforce or used subcontracted labour. A number of procurement, contractual and new financial arrangements emerged, reflecting different approaches to funding projects, managing risk and coordinating the production of increasingly complex products.

These changes were not solely the result of technological imperatives driven by requirements to install information and communication systems in buildings. Innovation in the production of buildings and structures was also stimulated by other changes in markets, by increasing national and international competition (often involving new players from supply industries) and by the introduction of more stringent regulations, especially those relating to the environment.

Construction firms came under scrutiny by building regulators concerned with environmental pollution because of waste created on sites, dust in the atmosphere, solvents in waterways and noise in cities. They were also targeted by environmental campaigners who objected to the destruction of natural habitats. By the 1990s, in engineering construction markets such as petrochemicals and nuclear decommissioning, environmental protesters had succeeded in gaining widespread international publicity for intervening in firms' practices. In road building, environmental campaigners had temporarily halted projects or contributed to their cancellation, proffering alternative technical and economic solutions. Construction was being forced to change its approach.

Neither traditional craft construction techniques nor the more recent industrialized methods from the machine age were adequate to cope with the new demands on construction. In Britain, the collapse of the Ronan Point tower block and lack of social acceptance of buildings produced from standardized heavy concrete components in the 1960s sent shockwaves through the design and construction community. Office buildings constructed in the 1960s and 1970s became obsolete because they could not accommodate information technologies and equipment needed in the digital age. At the same time, public sector construction projects had a poor record for completion on time and within budget. The level of public sector investment in buildings and structures also began to decline. Similar conditions were experienced across other OECD countries.

A modern approach to design, engineering and production was needed, including new ways of organizing building work to improve predictability and quality whilst reducing cost and construction times. However, despite internal and external pressures, design and construction practices appeared slow to change. Traditional methods continued to be practised in many projects, but leading design, engineering, construction and supply firms were proactive in developing innovative processes and technologies. A new production system was emerging, partly in response to the needs of innovative users and partly stimulated through competitive pressures on the supply side.

The digital age brought opportunities to use information and communication technologies in design and construction. This created further pressures on firms to reorganize processes in order to enable them to benefit from the use of computers in design and engineering and for construction logistics and project control. Greater emphasis was placed on improving the overall process: re-engineering design and production, partnering with clients and suppliers, and improving supply chain logistics. Some contractors became involved in the production and use of new types of pre-assembled components, in collaboration with manufacturers and suppliers. They also began to adopt 'lean construction' principles that emerged from Japanese manufacturing industries. Government policies in countries such as Britain and the USA supported these efforts.

The types of firms operating in the sector were changing, as were relationships between clients, those involved in the building process, and end users. New 'high-tech crafts' associated with the use of new materials and digital technologies emerged in specialist industries, whilst old trades continued to be eroded by the spread of industrialized construction techniques. But the pace of change was not evenly spread across OECD countries, nor was it experienced uniformly across construction markets. Japanese construction firms were dominant in terms of integrated processes and their ability to commit resources to research and development, whilst North American and European firms led in the use of information technologies.

Figure 6.1 illustrates the key driving forces and transformations in construction in the digital age. These are described in detail in this Chapter, which focuses on recent technological and organizational changes aimed at improving construction processes and products. The Chapter explores how new technologies have been developed to improve construction processes beginning with a review of general technical changes in construction materials and components. Many of these were based on improvements to standardized and pre-assembled technologies developed during the machine age. The Chapter continues with an analysis of information and communication technologies for design and construction processes. Achieving the benefits promised from information technology required changes in the organization of production. The new production system is the subject of Section 6.3, which includes a discussion of the changing role of design. The Chapter ends with an assessment of the performance of construction during the 1980s and 1990s, showing how new construction processes had emerged in response to the demands of the digital age.

6.1. Innovation in equipment, materials and components

A large number of changes occurred in construction process technologies during the first two decades of the digital age, many of which had their origins in the machine age and related to innovation in equipment, materials and components. For example, small handheld power tools, such as drills, saws and other cutting devices emerged and larger items of plant and equipment were improved through the addition of digital control and coordination systems. The mechanization of handling and lifting materials and demolition work in the 1960s brought the need for new skills and a new occupation on building sites — the mechanical equipment operator. Numerous types of cranes, hoists, trucks and mobile equipment were introduced and the range expanded with the development of smaller plant, for use

The driving forces coming from changing demand

Changes in work patterns

- Information technology use
- Flexible organizations

Changes in attitudes

- Environmental consciousness
- Health consciousness
- Rejection of standardization
- Drive for quality and lifecycle cost control

Changes in regulation

- Environmental
- Access for all
- From prescriptive to performance-based

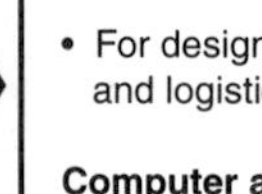

The re-invention of the construction industry

Interactive design

- Multiple agent participation for efficient high-quality client-adapted complex product with time dimension for flexible use

Flexible organization of work

- For design, engineering, production and logistics

Computer aided management

- Access to information for optimal decision-making
- Online processing:
  - Design and engineering
  - Operations
  - Coordination
  - Budget and finance
  - Project monitoring

The driving forces coming from changing supply

Availability of equipment and systems

- Generic information and communication technology (ICT) for users and producers (access to information, for data processing and transmission, etc.)
- Flexible portable ICT (for on-site use and intercoms)
- Programmable controls for construction tools

Availability of specialized software

- CAD: From 2-dimensional to 4-dimensional, for design, simulation, discussion, evaluation, decision making, coordination etc.
- Virtual prototyping and testing
- Lifecycle modelling

Materials and components

- Increasingly wider choice of materials and components
- Pre-assembled, tested and guaranteed components
- Possibilty of interaction with materials and component makers

Fig. 6.1. Construction in the digital age: driving forces and transformation

on sites with difficult access. Mechanization, which typically began on large civil engineering projects, diffused across the construction sector to include small-scale refurbishment work.

New types of fixing technologies were designed to ease on-site construction, resulting in a trend towards the development of standardized, universal joints, with the aim of providing quick-fit, clip-together assembly, rather than traditional shaping, honing and filling of parts on-site. The rationale was to provide right-first-time jointing with improved predictability of tolerances, drawing lessons directly from manufacturing. Examples included fixings for plumbing and pipework, electrical installations, steelwork, and composite panel cladding systems.

6.1.1. New materials

Construction processes consume large quantities of materials and components, supplied by a diverse range of industries, many of which have no relationship with one another except that their products may be combined together in the built form. The value of materials and components ranges from around 15 per cent of total costs in repair and maintenance to 70 per cent of the cost of producing complex intelligent buildings, with the average being around 45 per cent (Hillebrandt 1984, pp. 184, 253). The materials value chain typically involves both indirect and construction-related producers, together with those from the extractive industries, such as sand, stone and aggregates.

Drivers for innovation in materials differ considerably depending upon the type of materials firm, its relationships with other construction firms in the supply chain and the overall size of its market in construction. Many materials used in construction are produced for general application across a range of industries for which construction may form only a small part of total demand, for example, aluminium or plastics. Producers of these materials may not necessarily focus their attention on improving existing products or developing new ones specifically for construction, unless they have a particular construction products division. However, some materials, such as bricks and cement, are used almost exclusively by the construction industry. Technological developments by firms supplying such materials are likely to focus directly on improving products for construction applications.

The most important condition affecting the output of construction-related materials producers is the level of total construction output, but within this, the type and quantity of materials and components used depends upon the nature and composition of work. Some specialist materials and components are used predominantly in specific types of buildings, such as inert gases for fire-fighting systems used in computer and control rooms. Others, like concrete, are widely used in the production of almost all types of constructed products. In Britain in the late 1970s, at least half the output of most building materials and components producers was used in specific segments of the construction market, either in housing, civil engineering, industrial buildings, commercial buildings, or in repair and maintenance (NEDO 1978, p. 69). Some materials and components producers were therefore more sensitive to changes in specific construction markets than others and this also affected their approach to innovation.

Since the 1970s there have been substantial changes in both markets and technologies of building materials production. Some of these relate directly to the introduction of information and communication technologies in production processes. There have also been significant changes in the structure of the building materials supply sector, including a tendency towards increased international competition and concentration of activities in larger multinational businesses. In some OECD countries such as in Britain, this also coincided with a worsening trade deficit in building materials, particularly during the 1980s (Flanagan *et al.* 1995).

Innovation in basic materials originating from outside the construction industry has always played a major role in stimulating change within construction itself (Bowley 1960). Each successive phase of development in materials has been accompanied by their adoption in construction. By the 1950s, the chemicals and synthetic materials industries were developing new products that were to become particularly important sources of materials innovation for construction. By the 1960s and the dawn of the digital age, another materials revolution was underway. The roles of chemistry and materials science began to play a greater part in the design and production of materials for the built environment (O'Brien 1997, p. 66). Materials scientists were able to determine the mechanical behaviour of materials and thereby contribute to the work of engineers as much by telling them what to avoid as by finding new applications (Gordon 1976, pp. 258–259). New areas of technical specialism emerged, such as in protective coatings for metals and composites. These were either pre-applied in factories, or applied *in situ* on sites. They added an extra degree of complexity to construction processes because they needed to be protected from damage during other construction activities. Another emerging area was the use of synthetic materials such as geotextiles in geotechnical engineering.

Many new materials were adopted during the 1970s, including strong, rapid-hardening adhesives, flexible sealants and mastics, cleaning reagents, and synthetic materials such as plastics, PVC, polystyrene, Perspex and polyethylene (Freeman and Soete 1997, Chapter 5). Teflon (polytetraflouroethylene (PTFE)), synthesized by Plunkett at Du Pont in 1938, was gradually adopted in many different uses in construction, such as in plumbers' tape and, from the 1980s, for coatings used in lightweight tension fabric structures. Some products were developed for specific industries such as automobiles, shipbuilding or aerospace and were adopted later by construction. By the 1980s and 1990s, another phase of R&D in advanced materials resulted in the production of fine chemicals and new ceramics as well as recycled products. These materials began to find various applications in construction, from use in producing lightweight structures to insulation of buildings.

Producers of construction-related materials have been equally important in product and process innovations. Their proximity to construction activities meant that they were often able to combine new materials to form components and sub-systems for on-site assembly. This often involved the development of better materials and components aimed at improving construction processes, focusing on minimizing on-site work and improving the accuracy, speed and quality of construction. Examples included the use of new materials in structural framing systems, cladding and roofing systems and many internal elements ranging from partitioning, fixtures and fittings, kitchens, bathrooms, heating and ventilating systems, to electrical and control systems. These technologies were adopted either when they became cheaper or when they were perceived to offer clear technical advantages over traditional materials. Some were designed to enable the construction of lighter weight buildings, reducing the need for expensive foundation work and making erection easier. Others were designed to speed up construction by offering quick-fix components and rapid drying paints which required less skill in application whilst at the same time providing a greater range of choice in finishes.

The introduction of new synthetic materials stimulated innovation in older building materials industries. Even traditional materials such as bricks did not escape the pressures of technological innovation. For example, in Britain during the 1980s new investments were made to reduce cost and improve choice in brick production in order to maintain market share in the face of competition from new substitute materials. This resulted in new brick manufacturing capabilities including the introduction of computer-controlled equipment, facilitating a wider range of brick types and colours and helping producers to benefit from economies of scope and increased productivity. One brick factory, opened by Steetley in Britain in the 1980s had the capacity to produce 50 million reproduction handmade bricks per year, equivalent to 1 million bricks per year per employee—three times the number produced by hand in 1900.

By the 1990s, a major shift was occurring in materials science, the impact of which has yet to be fully realized in the development of new materials with potential applications in the production of the built environment. This was the shift from *discovery* to *design* of new materials. The most important plastic material, polyethylene, was discovered as the indirect result of other research (Freeman and Soete 1997, p. 123). Like many materials of the machine age it resulted from serendipity in experimentation. The big change came with the realization that new materials could be deliberately invented—designed for particular applications. Richard Feynman's famous 1959 lecture 'There's Plenty of Room at the Bottom' challenged scientists to manipulate and control matter on a very small scale, arranging individual atoms to make whatever was needed (Feynman 1960). In the 1980s the invention of the scanning tunnelling microscope, a computer imaging system with a surface probe, enabled the manipulation of atoms and molecules. Materials designers could begin to work at molecular and atomic levels, customizing specialist materials made to measure (Ball 1997). Nanotechnology had been invented and the digital age had provided the computer-aided tools needed to design specialist materials.

So, in the 1990s, those architects, designers, engineers and constructors who had been alert to developments in science were presented with the possibility of an array of potential uses in the built environment, ranging from new photonic materials to smart skins and materials for clean energy supply. Biochemical materials were being designed to assist in bioremediation of contaminated land and for cleaning processes. These had the potential to revolutionize areas such as civil engineering and geotechnics, for example, through the design of materials to lock-in pollutants. In other areas, new forms of durable concrete that stopped the migration of contaminants were being designed, together with fibre optic measuring devices to monitor pollution in the ground.

One consequence of the materials revolution was that construction professionals were inundated with choices arising from the potential offered by new materials. There were possibilities of using many of these alongside more traditional technologies, combined in pre-assembled component parts or applied on-site. This presented three problems.

(1) It was often difficult to make the choice of which material or component to use. One indicator of the growing range of parts can be found in the thickness of manufacturers' catalogues. In the 30 years from 1960 to 1990, this had increased tenfold in many cases, reflecting the increased range of specialist widgets available. This was partly the result of designers' practices, who often tampered with standard design details necessitating the use of bespoke parts. Manufacturers often kept these

modifications in their product ranges in case they were required for another project they appeared as additional 'specials' in subsequent catalogue editions. This created problems for future specifiers who had to cope with making choices between an increasing number of slightly different materials and parts that had been formulated to serve essentially the same purposes. For example, AMP, the world's largest producer of electrical connectors, makes 300 000 different component types. In 1998, it produced 300 new variants of connectors per day on behalf of its 200 000 customers in 125 countries. AMP was twentieth in the US league tables of companies filing patents for new products (Marsh 1998).

(2) New materials were often claimed by their manufacturers to be lighter, easier and quicker to manipulate on-site and to improve the aesthetics and physical durability of the final product. But questions remained about performance in use, particularly with respect to the relation of one material to another. Furthermore, performance in use was often difficult to simulate over the expected design life of many buildings.

(3) An increasing level of technical knowledge was required in using specialist materials on sites. For example, different paints, mastics and sealants were designed for use in specific climatic and environmental conditions: some could be used in elevations facing the Sun and were designed to resist degradation to exposure from sunlight, others were designed to withstand cold temperatures. Moreover, the simultaneous use of new and old materials created additional uncertainty over performance in use. New skills were needed to assess compatibility between materials, for example in the use of organic solvents. To complicate matters, these could sometimes be applied on-site by construction teams without verification by engineers or materials experts. Similar issues arose with the integration of component parts, including checking interconnections, tolerances and movement.

These issues themselves presented new market opportunities for some manufacturers of component systems who had the capability to integrate parts, test them and provide guarantees. By the 1990s, an increasing proportion of the value of projects was being added upstream in the supply chain by materials and components manufacturers who had invested in capital-intensive production processes and testing facilities. The role of component manufacturers and integrators grew, resulting in an increasing proportion of value being added further away from site-based processes.

6.1.2. Component systems

A brief history of the development of industrialized building components was traced in Chapter 2, showing a number of advantages of manufacturing component parts, moving work away from construction sites. Developments in the manufacture of standardized parts continued during the digital age, providing a growing range of pre-assembled components for use in all types of construction activity (CIRIA 1998). Many component innovations were dependent upon technological developments in other materials industries. For example, pre-assembled cladding systems and standardized factory-produced door and window units depended upon the use of sealants and mastics for their installation in buildings.

The impact of general technological changes in construction components and sub-systems varied across different construction markets. Some buildings and structures, such as office buildings, were constructed from a large range of parts. Others were designed from a large volume but small range of parts, for example many civil engineering structures. These characteristics of product types, together with the size of market, had a strong bearing on the possibilities of using standardized pre-assembled components. For instance, the design and construction of intelligent buildings generally involved more sophisticated approaches to systems integration than those used in housing, (with the possible exception of Japanese industrialized housing systems).

A number of distinct levels of componentization became evident. These ranged from modular or volumetric systems through flat-pack kits of parts and sub-assemblies to individual components. Volumetric systems included fully fitted rooms used, for example, in the production of MacDonalds fast-food stores, motels and student accommodation or in housing in Japan (e.g. buildings constructed by Toyota Homes and Sekisui Heim). Figure 6.2 illustrates production of a Sekisui modular house, the main advantages being speed of construction on-site and quality of the finished product. These methods have a long history through successive attempts to industrialize construction during the machine age. The modern variant of modular production was used extensively in the construction of off-shore oil platforms in the North Sea during the 1970s and early 1980s, in part to reduce the impact of adverse site conditions on construction times and quality. The success of these applications was influential in the decisions of architects such as Norman Foster and Richard Rogers who chose similar modules for bathrooms and plant rooms in the Hong Kong Bank and Lloyds Building in London. Similar approaches were adopted for the provision of complex plant and equipment such as air-handling units in many other intelligent buildings during the 1980s. Such approaches were also increasingly being used in a variety of new

Fig. 6.2. Sekisui Heim modular housing systems

and refurbishment projects, particularly where sophisticated machinery was required, for example, in refitting industrial kitchens in hotels, prisons, hospitals or events stadia. Figure 6.3

Fig. 6.3. Bathroom modules for the Hong Kong Bank and PKL modular refurbishment kitchen

shows a modular industrial kitchen unit, manufactured by PKL in Britain, being used in the refurbishment of student accommodation; it also illustrates the use of pre-assembled bathroom and equipment modules in the Hong Kong Bank. Whilst volumetric construction methods have advantages in terms of quality and speed of construction, disadvantages were often experienced in higher transportation costs and limitations on the extent to which modules could be customized to meet particular user requirements or to fit individual site constraints (Gann 1996).

Flat-pack component kits of construction parts and pre-assembled panel systems were widely used in many countries for walling systems, roofing and to provide interior fittings. Examples ranged from the cladding of office buildings to systems for housing, used extensively in Japan and some parts of Northern Europe (Gann 1999). Figure 6.4 illustrates pre-assembled walling systems used in Dutch and Japanese housing.

From an economic point of view, one of the main driving forces for the use of standardization and pre-assembly in the machine age was the desire to achieve economies of scale in

Fig. 6.4. Pre-assembled systems used in Dutch and Japanese housing

production. Manufacturing industries had demonstrated that, as the volume of production grew, cost per unit could be reduced more quickly than production costs increased. Component costs could therefore be reduced by standardizing stock items. This approach, however, had a number of disadvantages so far as the use of components in the production of buildings was concerned. The configuration of manufacturing technologies using expensive, dedicated machinery meant that large production runs were required and it was difficult to provide the variety in components sought by designers.

A partial solution to this problem came in the digital age with the advent of manufacturing processes using re-configurable machines and digital process controls. This increased flexibility on production lines and small batches could be produced to order for a particular project, rather than to stock in builders merchants. Firms were able to benefit from new economies of scope by developing processes that facilitated the production of a variety of parts using essentially the same machinery and material inputs, where previously different production lines would have been needed. These improvements in manufacturing resulted in the ability to exploit technical possibilities and deploy new capital equipment aimed at improving productivity and quality in the production of building components. However, realizing these possibilities in buildings required major changes in the organization of production.

To achieve the promised quality improvements and cost and time savings would require radical changes in overall design and construction processes. For example, architects and engineers would need to understand the implications of their designs for manufacture, assembly and systems integration, particularly with respect to the interconnection of parts. Construction managers would need to re-think their approach to the coordination of parts in allocating work packages for sub-assemblies to specialist suppliers. Some project processes were successfully re-engineered, resulting in changes in both the location of added value in supply chains and firms' traditional approaches to markets and competitiveness. These changes created new lines of responsibility and a greater need for control procedures to link construction sites and component manufacturers. Failure to change the organization of construction to accommodate the use of pre-assembled components diminished the likelihood of improving predictability in the quality of products or obtaining greater certainty over the control of processes.

In the 1990s, attempts to improve quality and meet the individual needs of customers were driven by consumer-oriented approaches in which quality and value for money were paramount. This necessitated thinking systemically about the organization of the total process and questioning where, at what level and to whom the use of standardization and pre-assembled components would be of most benefit. Other issues included the need to use simple connections and to consider product lifecycles and replacement costs. For example, there was usually a trade-off between the cost of holding spare parts and that of scrapping components completely and replacing them with new vintages. Few construction firms had been able to generate consistent data from enough projects to enable them to make detailed judgements about these issues. There were exceptions, however. Those best placed to understand the economic implications of using standardized component parts in a flexible manner included large Japanese industrialized housing producers and vertically integrated construction firms that incorporated manufacturing units, such as Skanska of Sweden. Others included firms specializing in particular markets such as the construction of housing, out-of-town retail stores, or motels. Evidence from Japanese industrialized housing production, where expert systems were used to coordinate millions of variables in component choice, indicated one way in which databases of standardized parts might be used more widely in the future (Bottom *et al.* 1994). The possibility of using information systems to coordinate the use of standardized parts promised to improve the capabilities of designers and production engineers in delivering customized solutions.

6.2. The impact of information and communication technologies

6.2.1. The promise

Of all the technological innovations in design and construction since the 1970s, the adoption of information and communication technologies (ICTs) has offered the greatest potential for improving production processes. The ability to manage knowledge and information effectively and efficiently has been central to performance improvement in many industries and ICTs have often formed the underpinning technologies upon which new processes have been built. An extraordinary array of developments in hardware, software and networks have emanated from the ICT industries, such that it has been difficult for user sectors like construction to keep pace with the possibilities from their introduction.

A number of general features characterize the potential benefits of these technologies, including the ability to automate information processing tasks leading to cost reductions, improved accuracy and faster response times. They also assist in providing new types of information about products and processes, enabling firms to establish markets for new services. These general features were discussed in Section 4.1.

There are two distinct, but often overlapping, ways in which ICTs have been used in the production of the built environment. First, they have been adopted to support project processes in design and management, providing mechanisms for linking decision making from early planning and conceptual stages through design, engineering and procurement to erection, installation, commissioning and even operation and facilities management. Second, ICTs have been used to assist firms in managing regular tasks relating to their internal business processes.

During the 1970s and 1980s, construction was a slow adopter of ICTs in comparison with many other sectors including general manufacturing and engineering (CICA 1990). By the 1990s, computers were to be found operating in almost every aspect of design, engineering and construction. There had been large variations in the use of these technologies by size and type of firm and by size and type of project. From the late 1970s to the 1990s, rates of adoption differed, with professional design and consulting engineering services firms in the vanguard (Soubra 1993).

Applications for computer aided design (CAD) were amongst the earliest to be used. In the 1960s, the ability to coordinate geometric information using computers was developed at MIT. This capability was essential for manipulating design, engineering, surveying and construction data. Programmes were initially highly specialized, running on large mainframe computers with applications oriented towards particular tasks such as designing steelwork, analysis of foundations, or planning and locating new buildings and roads. By the 1970s, general purpose CAD software became available, offering users significant advantages in handling general data for architectural and engineering design. The production of drawings and management of design changes were themselves major tasks on large construction projects—many tens of thousands of drawings and working details were often produced. Computers presented new opportunities to organize these tasks, saving time and reducing errors.

The specific circumstances found in design and construction projects meant that off-the-shelf software often had to be adapted to meet users' requirements. Leading firms established computing and information technology departments to assist with the introduction of these technologies. Some developed their own construction-specific software packages and a number of successful applications were sold to other firms. In some cases, construction computing departments became successful software vendors; for example, Ellstree Computing in the UK was a spin-off business from the John Laing construction company. During the 1970s and 1980s, other specialist software companies emerged and general software firms opened divisions to focus on supplying the needs of the construction industry.

The introduction of the personal computer in the early 1980s enabled the use of lower cost, more widely distributed CAD, from which many smaller design and engineering firms were able to benefit. Other applications were developed for modelling and simulating building performance, analysing and managing building costs and for project planning and management. By the late 1980s, cost reductions and performance improvements in hardware, together with better general purpose, easy-to-use software contributed to more widespread diffusion. The introduction of laptop computers in the 1990s provided additional benefits of portable computing to professionals who were often travelling to different construction sites or were site-based for short periods of time.

Telecommunication technologies were also developing rapidly during the 1970s and 1980s, providing new communications facilities that enabled improvements in voice telephony, including mobile communications. By the mid 1990s, this was revolutionizing construction activities, from the largest to the smallest projects. For example, even very small businesses owned by working proprietors were able to improve the coordination of their activities with other trades, materials purchases and delivery and waste removal, by using mobile telephones. New telecommunications systems also provided what became known as VANS (value-added-network-services). These enabled firms to link different digital communication functions across internal and external networks. Improvements in digital switching and higher bandwidth meant that it became possible to transmit large quantities of digital data rapidly, including drawings and virtual models, cost and project management data, and video images of construction sites.

Three enabling and applications technologies were particularly important in the development of ICT systems for design, engineering and construction.

(1) Databases were needed to store project information and provide easy access to different users at various stages in design, production and operation. The task of developing better ways to share data was important, particularly if ICTs were to provide potential benefits of improved accuracy and speed of information flow between different project participants. The concept of single project databases emerged. Instead of project data being held on separate databases by each firm, the idea was to produce an electronic data model to which all participants could refer throughout the processes of design, construction, operation and maintenance. During the early 1990s, this concept became widely accepted in the USA and Europe and was seen to be important for the integration of information flows. A number of large international industry and government research projects were

sponsored to develop database architectures, including the European COMBINE project. However, these projects experienced difficulties in achieving their goals because the nature of data requirements and types of users changed over project lifecycles.

(2) CAD software was useful in helping designers to coordinate basic geometric data, but it had limitations because it was not possible to manipulate data about other attributes, such as physical performance characteristics of component parts, shown in layered drawings. Software was needed that would enhance design and engineering activities by matching more closely the ways in which designers worked. This resulted in the concept of object-oriented applications in which each object (a part in a building or structure) had its own data attributes. An increasing range of types of information could be attached to objects and used in different ways. This could potentially enable the calculation of a wide range of cost and performance functions. As designs developed and were modified, it would therefore be possible to assess the consequences for each element, as well as to extract better information about the design as a whole. This technology had the potential to alter the ways in which information and data were manipulated, stored and transferred. However, in the late 1990s, a number of issues remained unresolved, including how to provide interoperability between applications requiring information from different object-oriented databases. While the concept was intuitively simple, the central idea that an object could carry with it embedded information about relationships with other objects made it difficult to implement. The use of such technologies was likely to require design using relatively standardized component parts.

(3) By the mid 1990s, use of the Internet for corporate communications was growing rapidly in the USA. By 1998 Britain and other European countries were also adopting Internet protocols for the transfer of business information, both internally within operating divisions and externally, between different suppliers and project partners. For example, in 1996, Bechtel developed Internet applications for transmitting virtual reality 3D images to clients in different countries. The range and types of available web browsers and access technologies was changing rapidly in the late 1990s and new encryption systems were improving data security. This offered the potential for project participants to use cheap and swift communications technologies to enable the

exchange of video, text and data. Internet technologies began to diffuse throughout construction.

The combination of high-powered portable computers and high-performance digital telecommunications provided the technological infrastructure which, in theory, had the potential to revolutionize project-based, one-off, design, engineering and production tasks. It could enable firms to improve performance in different activities organized in temporary, distributed networks. But to achieve these aims, hardware and software had to be developed to enable interoperability between different systems. This meant that standards were needed, and they became widely regarded by design, engineering and construction firms as essential if the adoption of ICT applications were to continue.

Standards for the exchange of drawings were particularly important. A number of commonly used *de facto* standards had emerged with the development of CAD, but as design and construction firms began to adopt 3D and 4D modelling tools in the mid 1990s, these *de facto* standards needed to be replaced by higher level alternatives with far greater semantic content. Attempts were made to develop new standards by software suppliers and by industry-wide standards-setting bodies. Firms in the USA were often the first to develop new standards, particularly when leading suppliers had targeted specific markets—for example, Autodesk's support of the Industry Alliance for Interoperability (IAI) standard or the extension of Microsoft's Object Linking and Embedding (OLE) standard to incorporate geometric properties, both of which aimed to enable software interoperability at levels that could display construction objects. Other standards set by industry-wide committees took a long time to finalize and were often superseded by technological developments: one example was the ISO STEP Standard 10303 which aimed to provide the underlying architecture for information exchange (Gann *et al.* 1996a, Chapter 3).

6.2.2. The reality

Investment in ICTs promised much, including improvements in the ability to provide structured feedback, through the systematic collection of information and use of databases, enabling learning within and between projects and project-based firms. The technology offered the possibility to improve inter-organizational networking, providing faster response times and more accurate information, including support for remote and concurrent decision making. It could also potentially assist in the integration of information flows, for example in the management of supply chains, saving time and reducing the number of defects and rework items caused by problems such as the use of out-of-date information.

By the early 1990s, major changes were occurring on large engineering projects where firms such as John Brown Engineering in the UK and Bechtel or H B Zachry in the USA were among the first to exploit the benefits of ICTs. This was particularly evident in work for the process plant and off-shore oil industries, where relatively few large international clients were able to articulate and enforce their requirements for performance improvements in design and construction. Markets for these types of engineering and construction facilities were serviced by international firms and projects were large enough to justify significant investment in ICT systems.

The rationale of firms investing in ICTs varied. Some viewed these technologies as enabling tools that could be used to enhance performance through information sharing in networks of interdependent suppliers. Others made investments for fear of losing out on potential benefits because their competitors were buying new systems, even when these had not been proven. In other cases, clients required project teams to comply with the use of particular information systems and sometimes firms purchased systems in order to demonstrate to their clients that they were using state-of-the-art technology. These conditions created a bandwagon effect that gradually gained momentum, such that by the mid 1990s, the majority of construction organizations were using ICTs in some, if not all, of their operations. Yet the slow rate of adoption and apparent inability of design and construction to derive benefits from ICTs indicated that the sector was not particularly successful in its investment in new process technologies. Moreover, evaluating and proving the benefits of these investments remained difficult.

The reality experienced by many firms was far removed from the benefits that had been hoped for. Investment and training costs had been higher than the perceived benefits. In many cases, 'islands of automation' had been achieved rather than fully integrated systems and therefore only modest improvements in productivity were gained. Figure 6.5 shows three phases in ICT adoption in relation to the business process changes required if the promised benefits were to be achieved. In many cases, firms had been unable to implement more than the simple substitution of CAD systems for more traditional approaches, such as using drawing boards. There appeared to be a general lack of understanding in all but a few firms of the full consequences of implementing IT strategies for the organizational structure of their businesses. In most cases, systems were introduced into traditional organizational structures that hindered the ability to achieve widespread benefits. This resulted in marginal trade-offs in cost, time or quality, rather than improvements in the overall process. Rarely were firms able to transform their business

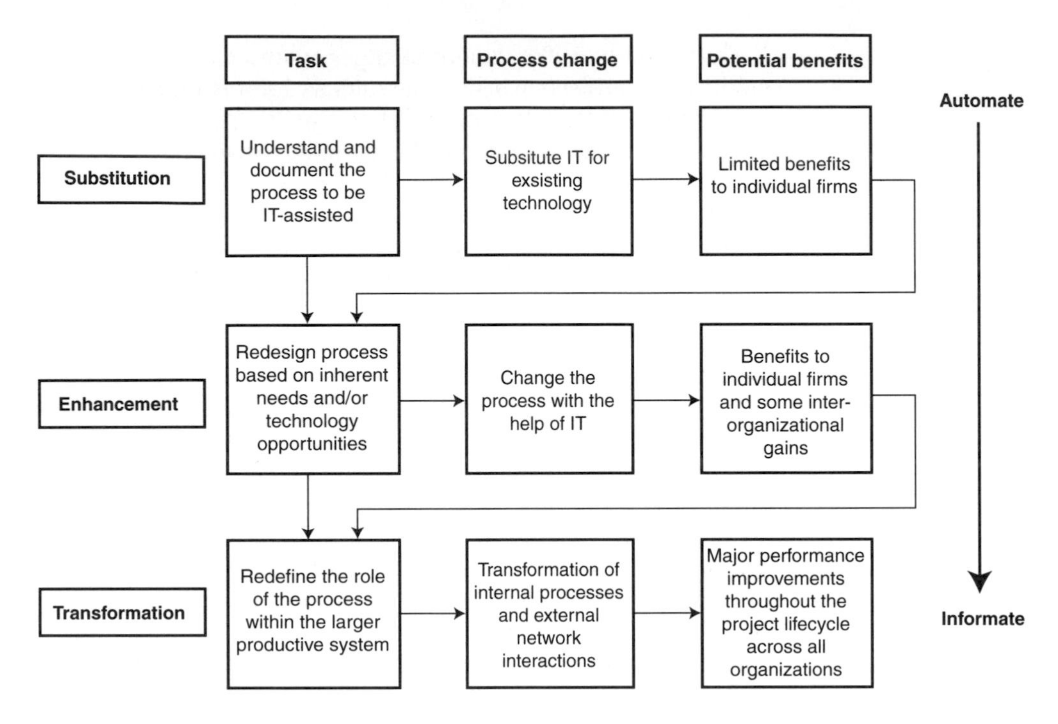

Fig. 6.5. Three phases of ICT adoption and business process change (source: adapted from Arnold and Gann (1995))

performance. Moreover, automation of some areas of work was hindered by bottlenecks in information flows elsewhere in projects. Incomplete and inconsistent datasets hampered the development of integrated systems and in many areas there was a lack of explicit, codified knowledge on which to base the development of ICT systems and databases. Many firms wanted to protect access to their own knowledge base and the level of trust between organizations was often too low for them to agree to use inter-organizational ICT networks.

This rather disappointing fulfilment of the promises of ICTs was evident in every field of industry, across most sectors of the economy and usually in the case of many users. Several studies by the OECD and others have attempted to explain why the expected productivity increases did not occur, despite the widespread adoption of ICTs (Roach 1991, Brynjolfsson and Hitt 1996, Morrison 1997). The expected benefits were either shown to be overrated, or failure to achieve them was explained as transitional growing pains. Much learning had to take place among producers and users before such a radically new technology was likely to reach optimal results. One consistent finding was the need to restructure organizations before they could effectively use new technologies and creatively adapt them to their specific needs and purposes.

A number of other general issues also hindered the rapid adoption of ICT systems in design and construction. For example, experienced senior staff were often reluctant to use new technologies for a variety of reasons and they often reverted to traditional methods. The proliferation of hardware and software systems, often with proprietary standards and protocols, made it difficult for many smaller firms to evaluate the strengths and weaknesses of competing technologies. The rapid rate of obsolescence of computing systems also made it difficult to decide upon which system would be suitable for future uses. This was particularly true in large projects that might remain active for a number of years. If the whole facility lifecycle was considered, it became almost impossible to predict the likely informational needs for operation and maintenance. For example, an enormous quantity of data is generated, collected and stored over a project's lifecycle. The duration of this data is usually far longer than the short-term and temporary nature of the coalitions of firms involved in its collection, manipulation and transfer. If it is to serve useful purposes it would also outlive several generations of ICT equipment. For these reasons it was difficult to introduce effective ICT systems designed to be used over complete project lifecycles.

In spite of these difficulties, ICTs were increasingly being used in two distinct ways: for product modelling and simulation (e.g. the representation of different attributes of building behaviour) and for decision support in design and construction processes with the aim of improving communications and control.

6.2.3. Product design, modelling and simulation

Computer aided design was the first major application of computers in the production of the built environment. Technologies were developed which often had close associations with traditional paper-based, document-centred ways of handling building information and this approach tended to dominate ideas about how to deploy ICTs. Moreover, CAD was often associated with drafting and the work of people whose background and training made them unlikely to want to work in new ways, such as adopting a database-centred approach (Voeller 1995). In the mid 1990s, many debates about the use of computers in design and construction remained focused on whether to use 2D or 3D CAD. However, some research teams in Europe and the USA believed that a new approach to building information was needed in order to focus attention on lifecycle design, and the integration of design, construction, operation and demolition phases. It was also thought that a new approach might enable greater exploitation of the increasing power of communication systems and computers, including the potential to simulate and visualize building attributes.

It was often impractical to build and test prototypes for large projects and the possibility of creating virtual reality (VR) models and other product simulations became attractive, particularly if these could provide tools for exploring alternative design and engineering choices. Research in universities and industry began to focus on project modelling and simulation technologies driven by the availability of low-cost powerful databases. New ways of representing building information were also being developed including the use of 'avatars'—digital representations of people in virtual environments (Groák 1998). Simulations of air movement using engineering tools such as computational fluid dynamics became commonplace and many other attributes—from acoustics to ground behaviour and soil mechanics—were modelled. A new digital design media had emerged (Mitchell 1995).

Digital simulations offered many advantages over traditional physical modelling, although they did not replace physical models completely. In some cases, this began to have an impact on construction processes, resulting in relationships with production teams in which designers were becoming more integrated within new construction processes. This was the case in the work of Frank Gehry in which the use of new technologies to support design decisions in creating complex geometrical forms radically altered design and construction processes, as described later in Section 6.3.

The advent of 4D digital design tools in the mid 1990s offered the possibility of better representations of building simulations through time. This also opened opportunities for new applications and methods of modelling, for example in space syntax—the study of space utilization. These techniques proved useful in making decisions in the early stages of projects, in client briefing, or involving different interest groups in planning. They could also be used to sign-off drawings with clients, remotely. For example, in the mid 1990s, Bechtel used VR 4D walk-through visualizations for remote decision making, linking designers in California with clients in Dubai, without either party needing to travel so often. These technologies proved useful in reducing risk and uncertainty, and improving predictability in design decisions. They also helped to lower travel and other costs associated with making design changes as well as saving time. More importantly, they illustrated the possibility of transforming client–designer–constructor relationships using VR visualization and auralization technologies. These transcended traditional approaches to displaying ideas and proposals, enabling clients and designers to think about them in fundamentally different ways.

6.2.4. Communications, decision support and process representation

The use of ICT systems to improve communications, decision making and organization of construction processes was another goal sought by leading design and construction firms. Work packages distributed across a wide range of different firms needed to be managed simultaneously. The supply of appropriate, accurate information to the right people when and where it was required became a critical factor for success. By the mid 1980s, the use of fax machines had become commonplace, speeding up the transfer of paper-based information. But whilst there were many attempts to use telecommunications for exchanging data electronically, this usually proved difficult to implement. Firms wanted assurance that decisions would be based upon common sets of data, eliminating problems experienced in translation and version control. Creating common databases and ensuring interoperability between communication systems was expensive, time consuming and often complicated by the interests of different systems suppliers and users.

Specific applications such as Electronic Data Interchange (EDI) and other electronic billing systems began to be used by some larger companies where they had longer term relationships with major suppliers. Suppliers and builders merchants introduced electronic stock control systems and in some cases on-site barcoded delivery and inventory systems were implemented. Smaller firms increasingly used personal computers for management and accounts purposes.

New high-speed digital communication networks offered the possibility of better links between clients, designers, construction organizations and suppliers. Some firms recognized that these systems had the potential to provide an infrastructure for knowledge acquisition and accumulation, enhancing the possibilities of providing feedback and the potential to learn from previous experience. Hitherto, conventional construction processes had usually involved sequential decision making, in which decisions were passed from one group of specialist professionals to the next, all of whom had to make inputs if overall project goals were to be realized. In this environment, decisions were translated and transferred in a process which changed semantics such that the original decision intent could easily be lost or altered. Multiple decision sequences of this type often took place with little or no possibility of assessing their validity against overall project objectives.

During the 1990s, firms in the USA appeared to be in the lead in most areas of development and implementation in design-, engineering- and construction-related ICT systems. A small group of firms participating in large and often international projects were able to procure and implement various ICT systems successfully. These firms had access to leading software suppliers and American firms in general were more advanced in their level of business networking and use of ICT systems than their counterparts in Europe or East Asia. Moreover, some of the world's leading research institutions were located in the USA and these worked in close collaboration with design and construction companies, providing a fertile environment for innovation (Gann *et al.* 1996a).

These early adopters of ICT systems recognized that forms of decision making and the nature of decisions would change if new technologies were implemented successfully. Evidence from implementation of ICT-based decision support systems in leading American construction organizations demonstrated that emerging construction processes were quite different in character from conventional approaches. Single project databases and new communications media, for instance, provided opportunities for virtually simultaneous decision making among project participants at long distance. These technologies introduced a new dimension to the integration of design and construction activities by changing the type of involvement of each participant, thereby altering the ways in which decisions were made. The timing, sequencing and hierarchy of decision making changed fundamentally. The most important aspects of change were:

- increased speed and concurrence of decision making;
- the potential to make information readily available when and where it was required;
- the possibility to increase the visibility of decision making processes, including access to other people's decisions.

The implementation of communication and decision support technologies also provided design, engineering and construction organizations with opportunities to carry out new types of work, offering customers so-called 'value-added' services with the intent of developing better user–producer relationships. Some design and engineering firms were able to extend their markets for services, moving into early project decision making and downstream facilities management. For example, R M Parsons, an American construction engineering firm, was able to offer clients new services based on its use of computer integrated project systems (CIPS) linked to geographical information systems (GIS). This was only achieved after the firm reorganized its business processes and relationships with suppliers and designers to enable it to offer one-stop project management services from initial inception to commissioning and operation of facilities, all supported by new ICT systems.

By the mid 1990s, interest in simulating production processes was growing, stimulated in part by attempts to re-engineer business processes. Firms like Ove Arup & Partners wished to model different processes in order to investigate ways of improving overall performance. Analysing processes depended upon the availability of suitable forms of representation and it was recognized that traditional methods of process representation in construction were inadequate, generically and for individual projects. Existing tools restricted manipulation of the potential process and the investigation of alternative processes. For example, they only provided limited means of representing the comprehensive range of activities and factors in design, engineering, procurement and assembly processes. At the same time, a variety of new methods of process representation based on computer simulation techniques were becoming available. The use of 4D visualizations and software produced by companies like SAP, BAAN and Origin to assist in 'Total Enterprise Management' could potentially be harnessed to develop better conceptual models of design and construction processes. By the late 1990s, developing new tools for manipulating processes remained a goal of researchers, software vendors and leading design and construction firms. Companies like Silicon Graphics were developing high-powered computing and visualization systems; others were developing expert systems and software agents to assist designers. Technological advances in visualization and database management systems therefore offered the basis for new forms of process simulation.

6.2.5. Instrumentation, control systems and automation

By the 1990s, digital technologies were being used in instrumentation, diagnostic equipment and non-destructive testing techniques across a range of construction design and engineering disciplines. The use of such techniques was becoming increasingly specialized, evolving into a new discipline in its own right. For example, handheld computers could be used to plug into building systems to provide fault diagnostics in much the same way as electronic car tuning systems. Moreover, these systems could be used remotely by specialists working in a different location from the building or structure being monitored.

New forms of instrumentation contributed to changes in the ways in which construction work was carried out and the monitoring, maintenance and adaptation of buildings and structures throughout their lifecycles. Systems were being devised to monitor the long-term durability of structures, facades, pumps, filters and fittings. For example, fibre optics could be buried in concrete under the ground to monitor stress and movement; sensors were placed in mechanical equipment to monitor flowrates and alert the need to clean filters.

Similar technologies were also being added to construction machinery, with the application of sensors and microelectronic programmable logic control systems to plant and equipment. The intention was to reduce the need for on-site labour, speed up work processes and improve accuracy. The use of mechanized systems on sites had implications for the ways in which buildings were designed and construction processes planned. Buildings needed to be designed to accommodate mechanized construction and this had consequences for the sequence and type of work carried out by a variety of trades. For example, by the mid 1990s, global positioning systems had been developed for use with piling rigs, by Stent Foundations in Britain. The aim was to improve accuracy in the positioning of piles in the ground, eliminating the need to use subcontractors to set out piles. This had major consequences for work organization and skills, including the need to train operatives and engineers to work more closely together through the use of digital communication systems and to relay details from computer-generated designs to piling rigs.

Another example involved the use of VR control systems for crane operators. Manipulating control levers in some types of cranes was a complicated task and it could take years for an operator to become proficient. By the mid 1990s, companies such as Bechtel in the USA were experimenting with the use of data-glove control devices to enable crane operators to learn and operate machines more efficiently. Similar systems were also being developed to improve safety of lifting operations. The use of laser levelling devices connected to programmable excavators or the use of programmable concrete screeding machines illustrates other examples of the ways in which digital control systems were being used on construction sites.

The development of construction robotics represented the most advanced form of automation. In the early 1990s, Japanese companies led the world in developing and implementing these technologies with their strategy of adding programmable controllers to fairly standard items of equipment, including radio-controlled power floats, ceiling panel erectors and inspection robots (Gann and Senker 1993). Major Japanese construction firms (such as Taisei, Obayashi and Shimizu) attempted to develop on-site automated factories during the late 1980s at the peak of the construction boom. These prototype factories were located on the top of the structure to be built, providing a covered environment for work to proceed below. Robots were used to construct the structure and cladding of the building and as each floor was completed the factory was jacked up to the next level, making way for construction teams to complete interior works below. Figure 6.6 shows an illustration of Obayashi's automated building construction system (ABCS).

These automated factories had limited use because they were only effective on large standardized building types. Taisei Corporation's T-UP system was used successfully in the construction of a 34-storey headquarters for Mitsubishi in which

Fig. 6.6. Obayashi's automated building construction system

steel erection work was achieved at a rate of one floor every three days. However, the overall construction time of 24 months was no quicker than many conventional methods. The slump in construction activity in Japan by the early 1990s resulted in a slow down in development work on these technologies. They had not proved beneficial in terms of cost and time savings, however the experiments were useful in developing better knowledge about the viability of automating on-site construction work. Smaller scale, specialized programmable tools and equipment were proving more useful. Moreover, increasing the level of mechanization had implications for skills across the industry from designers and managers to machine operators and new equipment maintenance skills, including the need for knowledge of programmable machines.

6.3. New ways of organizing processes

New ways of organizing design and construction processes emerged after the Second World War in the USA, UK and across Europe, focusing on the need to manage increasing levels of technical complexity in projects. The need to improve project management processes was growing in importance, particularly in the defence industries. By the 1950s, a new form of project management, known as systems engineering, had been devised to improve delivery of complex military projects such as Atlas, the intercontinental ballistic missile system in the USA. These military projects involved tens of thousands of firms and hundreds of thousands of people working on complex technical problems. Modern project management used in producing large systems was born and spread across the civil sector to include large building and infrastructural projects (Morris 1994).

By the digital age, managing large construction projects usually involved dealing with complex social and political issues such as the need to include the concerns of a wide range of environmental and other interest groups. Many infrastructure projects in Britain were halted or delayed by the actions of environmental pressure groups during the 1980s and 1990s, including road building for the Twyford Down and Newbury bypasses and Manchester airport's second runway. Large housing, office and retail developments were also challenged. Project management approaches developed to manage technical complexity had few answers to these new social and environmental concerns. Rational, systems engineering approaches of the 1960s failed to provide the tools for managing in a world in which increasing levels of social, political, economic and technical complexity had to be accommodated (Hughes 1998).

By the 1990s, new, more flexible and adaptable management approaches were emerging. Rather than treating environmental concerns and other local community interests as externalities, some projects were planned to include representatives from different interest groups from the beginning. A new style of participative and open project management was evolving. This combined the rational logic of the previous era with adaptable approaches, enabling plans to be changed to accommodate local requirements. For example, instead of relying solely upon technical specialists to solve mechanical, electrical and controls problems in the construction of intelligent buildings, new styles of management involved interdisciplinary teams, including environmental engineers and social scientists. This approach was evident on the Boston Central Artery/Tunnel, the largest civil engineering project in the USA during the 1990s. The project was managed by a joint venture between Bechtel and Parsons Brinckerhoff, two of the world's largest consulting engineering firms (Hughes 1998). This team attempted to deal with an increasing number and range of complex issues on a project of huge scale and public visibility, where it was seemingly impossible to avoid conflict and delays. This approach was indicative of a change in management styles from the Modern to the Post-Modern eras, roughly corresponding to the machine and digital ages, as shown in Table 6.1.

Resolving complex technical issues remained important and the installation of digital technologies, together with the use of new materials and component systems, had a major impact on the ways in which buildings were designed and constructed. But organizational changes were needed to accommodate growing

Table 6.1. Characteristics of technology and management in the machine and digital ages (source: adapted from Hughes (1998, p. 305))

Machine age	Digital age
Production system	Project
Product delivery	Bundled product and service delivery
Hierarchical/vertical	Flat/layered/horizontal
Specialization	Interdisciplinary
Integration	Coordination
Rational order	Complexity
Standardization/homogeneity	Heterogeneity
Centralized control	Distributed control
Programmed control	Feedback control
Manufacturing firm	Joint venture/project-based firm
Experts	Meritocracy
Tightly coupled system	Networked system
Unchanging	Continuous change
Hierarchical decision making	Consensus-reaching
Seamless web	Network with nodes
Bureaucratic structure	Collegial community
Incremental	Discontinuous
Closed	Open

numbers of technical specialists, environmental experts, economists, ergonomists and other social scientists at the early stages of planning and executing complex projects. These changes were fostered by innovative construction firms, often working closely with a few knowledgeable clients. They included new forms of contractual relationships and supply chain management techniques that were to prove as important as technical change in improving overall construction performance. The potential to exploit new markets together with competition both from within and outside the traditional construction system resulted in a climate of rapid innovation in technologies and in the organization of construction processes. New players moved to centre stage in the construction system, resulting in changes in competition and industrial structure. In Britain, the sector became more fragmented and specialized; the number of small firms grew and large firms diminished. A variety of different types of specialist subcontractors filled the space left by the demise of traditional general building contractors. The role of mechanical, electrical, controls and data cabling specialists grew in prominence as the value of their work increased as a proportion of total construction output. However, they did not—or were unable to—take advantage of their new roles in relation to the management of the whole process.

In Japan, leading construction firms remained vertically integrated and they played an important role as innovators, by developing materials, component systems and construction

technologies themselves. But in most countries only a few design and construction organizations had the capabilities to participate in developing technologies, although many responded to the requirements of producing intelligent buildings by making changes in the ways construction processes were organized. For example, general contractors sought to change their market positions by developing and offering different services, shifting away from direct ownership and control of construction work to providing business services in a variety of forms of construction and project management. They coordinated specific erection, assembly and installation tasks carried out by subcontractors. Many other firms resisted changes and faced increasing competition from those with innovative responses.

An array of design and construction processes and contractual conditions emerged in the 20 years from 1970 to 1990. Methods of organizing construction varied significantly across OECD countries and these had consequences for the development and introduction of new technologies. Project organization varied depending upon the types of firms involved and sizes of projects they were working on. The organization of many small projects remained rooted in indigenous craft methods. Larger projects tended to exhibit modern international engineering and assembly approaches. These changes were exemplified by the proliferation of different methods used for procuring building work in Britain, as clients and construction firms struggled to find better ways of buying and selling.

A number of problems had become evident in traditional forms of contract used in the machine age. For example, divisions between conceptual stages of design and engineering and practical stages of construction resulted in a linear process, giving rise to long project times and it was difficult for construction teams to learn from their experiences when each project was divided into separate contracts. Legal and institutional frameworks often hindered the ability to plan for effective use of resources and develop new technologies. Moreover, those responsible for carrying out construction work were often excluded from providing advice on how to design better processes to minimize time and costs and improve quality because they were not involved until after the design stages had been completed.

Many clients had little knowledge of construction processes and were incapable of intervening in problems caused by the separation of design and construction, or by factional interests of different organizations in the production system. Clients often bore the cost of problems inherent in traditional contracting such as extensive project delays, prolonged litigation actions, poor quality control and slow, piecemeal innovation. Such problems were most pronounced on large, complex engineering-type projects. Clients, contractors and others involved in the construction process responded by developing alternative procurement methods and new forms of contracting. In Britain, during the latter stages of the machine age there were three main forms of contract used in public works, private building and civil engineering. By the mid 1980s dozens of different types of contract were in use.

Three generic types of project organisation were identifiable in different regions. They coincided with different styles of construction work in the USA and UK, northern Europe, and Japan as explained below.

(1) First, the Anglo-Saxon approach found in the USA and UK was typified by unstable price-based market transactions where temporary coalitions of firms were formed to deliver one-off projects. This form of organizing construction resulted in adversarial relationships in which risk was managed in hierarchical networks of firms. Firms remained independent from one another and often treated specialist project information as a strategic asset, concealing information rather than sharing it. Information flows were often disrupted and it was difficult to enable work to be completed smoothly. This caused uncertainty over project outcomes. Under these conditions, firms found it difficult to commit to long-term technical development and R&D, but the production system was highly flexible and enabled firms to respond rapidly to cyclical fluctuations in markets. In the 1970s, the British general contracting firm Bovis, which had pioneered a form of contracting known as 'fee management' in the late 1920s, promoted new producer–user relationships with clients. This approach, known as 'management contracting', involved the supply of business services required in arranging, coordinating and managing construction projects for a fixed fee. It became popular during the recession of 1979 to 1982, when main contractors shed many of their directly employed workers. Technical changes resulting in increased specialization meant that specialist skilled and semi-skilled labour was often required more than traditional craft trades, but specialists were only needed for relatively short periods of time on any particular project. Prime contractors preferred to manage subcontractors rather than bear the risks of employment directly. Management contracting and its variants provided them with the immediate short-term flexibility they required to respond to different client needs. Other approaches to

providing flexibility in coordination and the management of risk in complex projects also emerged, including 'design-build' and 'construction management'.

These approaches aimed to clarify the allocation of decision making responsibilities between clients, designers, main contractors and subcontractors, and thereby ensure that complex projects, such as the construction of intelligent buildings, were more manageable and were completed on time and within budget. These new arrangements sought to assist managers in off-setting risks through pyramids of subcontractors. New forms of contract often focused on dealing with increasing numbers of disputes arising from passing liability down the supply chain. Such approaches received sharp criticism from many subcontractors and suppliers who found themselves carrying more of the risks for performance to time and budget, in an increasingly competitive environment.

(2) A second general approach could be found in Japan, East Asia and some parts of Europe. It involved relatively stable partnerships between firms. These enabled long-term inter-organizational networks to flourish, facilitating integrated decision making capabilities across project teams. This created environments in which firms could transfer technology and collaborate in longer term R&D. Examples included the types of relationships found in Japanese construction. But whilst this had merits over the Anglo-Saxon approach in that it fostered technical development and information sharing, it was relatively inflexible, particularly in response to rapid changes in demand.

(3) The third general approach to organizing construction was evident in Scandinavian and other northern European countries. This involved strategic partnerships between clients and producers and between contractors and their suppliers. In the 1990s a hybrid, partnering form of organizing construction also emerged in the UK and USA in response to the need to deliver projects on time, within budget and to the required quality, without immediate recourse to litigation whenever problems arose (Barlow *et al.* 1997). This approach enabled the transfer and development of core technical capabilities, particularly on North Sea engineering and construction projects. It provided incentives for sharing information and collaboration in technical development and it was also more flexible than the Japanese model.

New contractual arrangements and ways of organizing construction were closely linked to changes in project management. Ideas for managing systems integration in complex projects originated in the USA, particularly in the defence sector. In the early 1970s, firms such as Bechtel developed methods for managing and reducing costs in the construction of petrochemical plant and oil installations in the Middle East. These project management methods usually involved single-point responsibility on 'turnkey' projects in which clients employed firms to coordinate and manage complete projects from design through construction to handover of finished buildings and facilities.

In the 1980s, contractors developed management techniques to cope with the need to control concurrent activities on sites. 'Fast-track' construction was the generic term used to describe an approach to managing parallel processes aimed at cutting costs by speeding up construction work, giving faster capital turnover time and thus shortening the time that clients' capital was tied up in production. It was in some respects similar to Japanese 'just-in-time' methods used in manufacturing (see Schonberger 1982), but it owed its origins to techniques developed by American project management teams. In North America during the 1980s, fast-track became an ideology for instilling performance criteria into the workforce. In Britain it meant something more specific: it was a method for overlapping design and construction in concurrent processes. Work was divided into packages that enabled working drawings to be prepared just-in-time for construction work to commence, rather than several months or even years in advance.

On large projects, work was usually carried out in phases, the idea being that this permitted parts of construction to begin before all the design stages were complete, thus reducing overall project times. It also permitted phased occupation before final project completion, therefore providing clients with an earlier return on investment. The potential time saving due to concurrent design and construction in fast-track approaches is illustrated schematically in Fig. 6.7.

In Britain, fast-track methods were promoted by proactive clients (such as the property developer Stanhope) who were prepared to enforce punitive measures to ensure that all those involved in the supply chain performed satisfactorily. Stocks on sites were kept to a minimum. Modular components were installed in their final location immediately on delivery, whenever possible.

One of the best examples of fast-track management was in the construction of a large speculative office development called Broadgate, at London's Liverpool Street Station in the mid 1980s. The plan was to build 3.25 million square feet of office space to accommodate up to 30 000 office workers. The size of this project added weight to the developer's (Rosehaugh–Stanhope) threat to use overseas construction expertise and

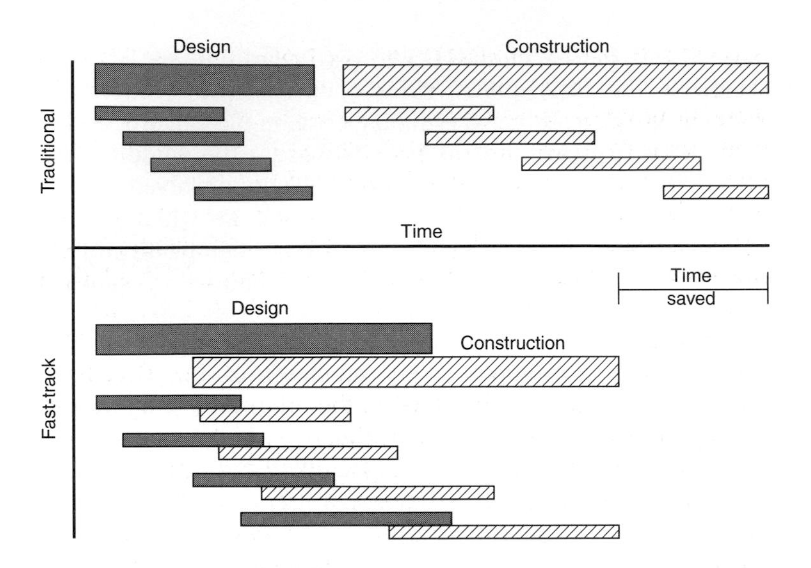

Fig. 6.7. Potential time saving due to concurrent design and construction in the 'fast-track' approach (source: adapted from Fazio et al. (1988, p. 196, Figure 1))

materials if British firms were unable to adopt new approaches. The management contractor, Bovis–Schal, an Anglo-American consortium, divided the work into 14 phases, with a design brief for shell and core offices to be fitted out by tenants. Detailed design work was carried out while initial construction work was underway, and design for later phases was undertaken while earlier phases were under construction. Work began in 1985, and phase one was completed within one year. Six months later, tenants had finished fitting out their interiors and they had moved into occupation. The total design to handover time for this 1 million square foot phase was around 23 months; this compared with between 32 and 36 months for a conventionally built office building, representing a 33 per cent saving in time.

Fast-track was primarily a management technique for speeding up work, but it was also closely linked with changes in technology. Management teams aimed to reduce on-site labour in order to achieve faster construction times. Shortages of skilled labour in Britain during the 1980s were an additional incentive for reducing reliance on traditional craft trades. This resulted in the use of prefabricated components wherever possible, for example, in the form of structural elements, toilet and bathroom pods, modular plant rooms, lifts and cladding panels which included radiators as well as ductwork and piping. Modules often weighed several tons, and were floated into place on air-skates. Victaulic joints for prefabricated pipe runs were developed, permitting connections to be made ten times faster than those with traditional welded joints. New lifting equipment was used to move components and site management teams used computerized management systems to schedule and monitor work in progress. On-site construction processes had changed from sequential activities associated with working on basic materials to concurrent assembly of component parts.

The Broadgate development was in many ways a showcase for management contracting and fast-track construction. It was run by a competent management team who maintained tight control over the project. The approach was adopted on a number of other large projects in and around London during the 1980s and 1990s. For example, many members of the Broadgate management team were involved in the office redevelopment of Canary Wharf in London Docklands, which dwarfed the Broadgate project in terms of size. The use of fast-track methods was not limited to the construction of office buildings. It was also used in housing, such as in the construction of Cascades, a 20-storey block of luxury flats in London Docklands. The developer, Kentish Homes, completed this project in 18 months using prefabricated modular components, and the first residents moved into the lower floors while heavy construction work was continuing above their heads. However, this attempt at early, phased occupation resulted in a number of disasters, including water leaking into completed flats, caused by poor workmanship and management.

The success of fast-track therefore depended upon highly skilled teams who could attend quickly to design errors, omissions and changes, and at the same time coordinate design, phased construction and work packages. Successful implementation required skilful individuals capable of responding quickly, without necessarily relying upon support from other managers in their firms. Inadequate skills resulted in disruptions within a complex of mutually dependent parallel operations, and there were fewer mechanisms through which other professionals could step in and sort out the ensuing chaos. Unexpected additional costs sometimes resulted, and the techniques did not always result in shorter project duration (Fazio *et al.* 1988). Attempts to introduce just-in-time design meant that it was difficult to alter design decisions and there was an inherent contradiction in the use of prefabricated elements which required early design, not just-in-time design, in order to allow factories time to plan production of components. Moreover, mistakes sometimes had more severe consequences than in traditional approaches, because the decision making time was compressed. In traditional sequential contracting, independent professionals monitored work and the hierarchical command chains within main contractors' firms, from building sites back to head offices, provided additional expertise if required. When something went wrong in the traditional sequential mode of operation, other people in the management structure were often able to intervene effectively, although this took time.

By 1990, work on several large sites in London, where these approaches had been adopted, was badly behind schedule. This was partly due to problems of control and the need to rely upon skilled managers and operatives. During boom periods it was possible to motivate people to work faster through various incentive schemes such as bonus payments because there was always a new project for them to move on to. Employment was assured, although firms phased pressures of wage inflation and the risk of staff being poached by other firms. Once a downturn was perceived, as was the case in 1990, people were inclined to 'soldier' or go slow, because they did not wish to hasten the time at which they became unemployed. Achieving the aims of fast construction therefore appeared to be more difficult during recession. Furthermore, it was possible that new methods of procurement led to escalations in costs because it involved many firms bidding for smaller work packages. Hundreds of tenders had to be drawn up for each project. When tenders were drawn up for both detailed design and construction, as was increasingly the case in many specialist areas, the cost of tendering was high, and the costs of unsuccessful tenders were eventually passed on to clients in future projects.

New approaches to managing technical, social, financial, political and organizational complexity appeared to have failed, particularly when measured against the need to deliver new types of buildings and structures whilst reducing costs and speeding up processes. These new approaches often resulted in disputes and litigation. In spite of success on some projects, they had not provided a solution to the problems of managing the design and integration of complex building systems. During the 1990s in Britain, two attempts were made to diagnose the root cause of these problems. The first involved a review of construction practices by Sir Michael Latham, concluding that new forms of partnering would help to reduce the adversarial nature of construction and improve information flows between different participants (Latham 1994).

The Latham Review, published in 1994, was followed by the Deputy Prime Minister's 'Egan' Construction Taskforce Report, published in 1998 (Construction Taskforce 1998). This report endorsed the partnering approach and also laid emphasis on the need to develop mechanisms for continuous improvement, including setting annual targets. The Taskforce recommended new approaches to design and product development, as well as making more effective use of standardized and pre-assembled components, the collection and use of better performance metrics and improvements in the flow of information between participants. This work drew heavily from lessons in Japanese 'lean production' methods and from a growing body of academic work on the benefits of 'lean construction' (see Womack *et al.* 1990, Womack and Jones 1996, Alarcon 1997). By the late 1990s, different approaches to value management, concurrent engineering, just-in-time delivery, waste reduction and business process re-engineering were being trialled in Britain, North America, parts of Europe and Australia, whilst many Japanese firms were already using some of these methods.

In 1999 it was too early to establish the overall success of these initiatives, except on isolated projects where improvements were being measured in detail, using new key performance indicators. In these cases, improvements appeared to have been made. However, demonstration projects in the UK showed the necessity to move away from traditional linear–sequential models of design and construction towards a new adaptable, interactive and integrated approach, shown schematically in Fig. 6.8. This approach would involve feedback from users, as well as those involved up and down the supply chain, creating a network of organizations that could learn from previous projects. It could provide an environment within which contractors could continuously improve by reducing defects, materials wastage and time and cost overruns, thus delivering better value to their customers.

6.3.1. Design

New ways of organizing production and the introduction of information and communication systems had profound consequences for design activities. The traditional role of architects as leaders of construction projects was in many cases eroded. In the machine age, the decoupling of design from production had

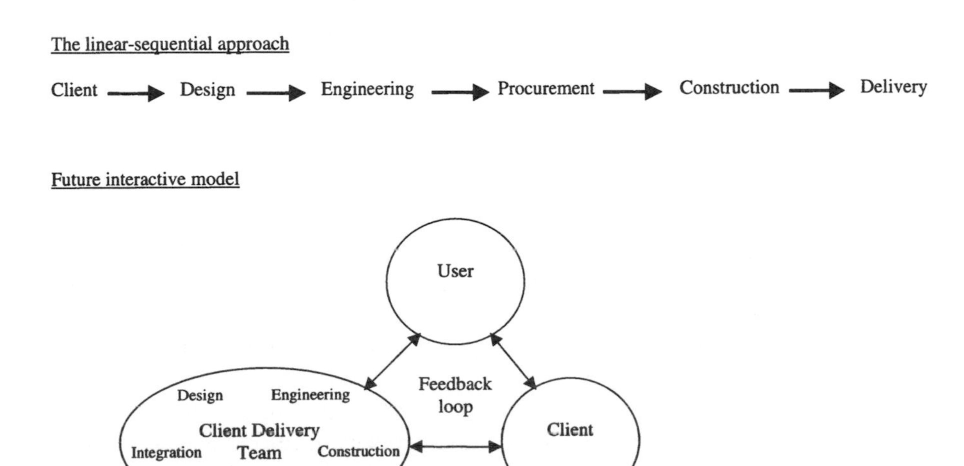

Fig. 6.8. From a linear–sequential approach to an interactive–integrated model of construction

meant that, in many countries, architecture remained largely within the sphere of the fine arts while construction work was regulated by the economics of production. The artist's dream had to be realized in a world conditioned by the need to reduce costs and introduce new technologies. This often resulted in tensions between the professional roles of designers and the commercial imperatives of contractors. Architects were often responsible for liaising with clients, although clients' expertise at dealing with the architectural profession varied enormously. Many clients had little knowledge of construction processes and they were incapable of intervening to resolve problems caused by the separation between design and construction and by factional interests of different organizations in the production system. Increasing technical complexity meant that detailed design work inevitably migrated up the supply chain into manufacturing firms who had the capabilities to engineer and test standardized, factory-produced component parts. Design activities were dividing into routine processes in which ICTs could be more easily deployed, and creative processes where the role of intuition and tacit knowledge remained important.

Vitruvius' three principles of firmness, commodity and delight had guided and challenged architects and building designers for nearly two thousand years. These principles were accepted as goals to which building designers aspired. But the ability of architects to maintain control over the total process and pay attention to design details was eroded as technological systems became more complex in the digital age. The design of intelligent buildings required new skills in understanding user requirements and integrating sophisticated technological systems. In this respect, design work was becoming more important because it was through the capture of user requirements, consultation, briefing, planning and good schematic design that value could be added during the early stages of projects.

Some designers were able to exploit new opportunities by using technology and reorganizing design and construction processes. For example, architects (such as Frank Gehry, discussed later), were able to use ICT systems to facilitate the involvement of designers in the *total process*. This increased designers' responsibilities in overall construction processes. Standardization of component parts and repetition of processes could be achieved—improving buildability without sacrificing form.

By the 1990s, design processes were changing rapidly as were the roles of design and engineering professionals. In many areas, architectural design appeared to have become increasingly divorced from systems design and engineering. There were two related reasons for this. First, the complexity of buildings and structures had increased with the introduction of new technologies and the need to accommodate flexible patterns of use, whilst reducing environmental impact. Specialist roles had emerged in response to the need for new knowledge in these areas. Second, the use of information and communication technologies in design processes was changing the ways in which designers worked, providing possibilities to simulate and test design options and coordinate activities between different specialist designers.

Ideas about building design had also changed with the rise of the digital age. Innovation in architectural design during the machine age was dominated by Modernist ideas. Large, centrally planned projects were conceived by architects and planners who supposedly represented the community's best interests and could reflect what Mies van der Rohe called the 'will of the age'. The Modern Movement was, however, challenged during the 1960s and faith in its abilities to deliver socially and politically acceptable architecture was eroded. It was during the 1960s that the anti-Modernist movements sprang into life. There was growing public discontent with the role played by technologists, bureaucrats and rational forms of decision making in monolithic state institutions, which often determined what and how buildings were produced. By the early 1970s, the consensus on what and how to build was broken and the leading roles and power enjoyed by architects and planners was opposed through the expression of new ideas about the role of design. Some wanted to turn the clock back and return to traditional vernacular styles, craft values and production methods. In Britain, Prince Charles stimulated the architectural debate, arguing vociferously for a return to traditional values, using various terms to describe what he disliked about modern architecture. For example, he called successive buildings designed in the 1970s and 1980s a 'glass stump', 'Victorian prison', 'nuclear power station', 'hardened missile silo', 'broken 1930s wireless set', 'grubby laundrette' and 'monstrous carbuncle'.

Other ideas were gaining ground, taken eclectically from the past and present. There were calls for 'pluralist', 'organic' solutions rather than the totalitarian view of the central plan. Post-Modernism had arrived. Good design practices were revised to accommodate particular social, economic and political concerns. For example, during the energy crises of the 1970s, the Royal Institute of British Architects promoted the 3Ls concept of Long-life, Loose-fit and Low-energy, aimed to help designers meet demands for low-energy buildings and improve flexibility in use. The transition from the Modernist Machine Age to the Post-Modern Digital Age was marked by a change in architectural styles, described by David Harvey (1989, p. 40):

The glass towers, concrete blocks, and steel slabs that seemed set fair to steamroller over every urban landscape from Paris to Tokyo and from Rio to Montreal, denouncing all ornament as crime, all individualism as sentimentality, all romanticism as kitsch, have progressively given way to ornamented tower blocks, imitation mediaeval squares and fishing villages, custom-designed or vernacular housing, renovated factories and warehouses, and rehabilitated landscapes of all kinds, all in the name of procuring some more 'satisfying' urban environment.

Yet Modernist designers had taken an active interest in how their buildings were produced, often intervening in production and designing new processes. In the 1970s, 80s and 90s, British architects such as Norman Foster, Nick Grimshaw and Richard Rogers continued this tradition, pressing ahead with the use of new technologies, exploring different ways of improving products and processes. But many contemporary architects appeared to play a lesser role, being interested in the images they created rather than the means by which buildings were produced, reinforcing the divide between design and production and the role of specialists in their own domain. This was in part reflected in a diminishing role for architects in the production of the built environment. Post-Modern architects were able to develop their pastiche from a wide palate of readily available materials. The plentiful supply of skills required to build great cathedrals, castles and mansions had disappeared long ago. Ornamentation could now be reproduced synthetically using new processes and materials. Dispersed and decentralized urban forms had become more technologically feasible than at any time during the machine age (Jencks 1988). Herein lay the irony of Post-Modern architecture: that it could only be achieved economically through the use of modern production techniques and materials. The use of new design technologies, such as computer modelling, had diminished the need to produce repetitive, standardized architecture from standardized components. In spite of a period of rapid innovation in materials and construction processes, many Post-Modern buildings were designed to give the appearance of not resulting from highly rationalized production processes. Their function was often concealed, in stark contrast to buildings of the previous era where form and function were united and exposed. Almost any style could now be achieved.

Yet some architects attempted to unite product and process design. One of the best contemporary examples of linking technology with new architectural design and production processes can be found in the work of Frank O. Gehry. His sculpted, gesture-based, building forms would have been technically difficult, if not impossible, to realize before the late

1980s. Gehry's best known works include the Guggenheim Museum in Bilbao, Spain (1991–1997), the 'Ginger and Fred' office building in Prague (1992–1996) and the Cleveland Weatherhead School of Management (1999). These buildings satisfied Vetruvius' requirements for commodity and certainly for delight, even if they may appear light and ephemeral when it came to the test of firmness.

Gehry's work was based on highly sculpted, geometrically complex forms, created through an iterative process of physical and computer modelling. The curved surfaces in geometrically complex designs were difficult to model using traditional CAD systems. In 1990, Frank O. Gehry's company invested in a new design and manufacturing system, known as CATIA, developed by Dassault Aerospace, having observed its use in producing curved shapes at Chrysler automobiles and in the design of Boeing's 777 aircraft. This system was introduced as part of a radically new approach to design, development and construction, managed by James Glymph, an expert in production process design. The process created the opportunity for Gehry to be involved in more of the overall design process than had previously been possible. He produced the physical, sculpted shapes by hand. From these, scale models were produced by Gehry's assistants and scanned into 3D computer models using a digitizer (a medical plotter originally designed to record the shape of the human head for brain surgery). The digitized forms were then manipulated using CATIA software on IBM RISC 6000 machines, providing the ability to model and engineer every spline and node point accurately. This was an advance on the polygonal rendering techniques used in most CAD systems. More physical models were then created for design verification using rapid prototyping machines, including the Helysis solid modeller. Once the final design had been agreed, electronic design data could be transferred digitally to various specialist fabricators working in steel, stone, glass, titanium and composite materials. The design office worked closely, although 'remotely', with suppliers and subcontractors to reduce costs and improve buildability.

This process was successfully used to coordinate the production of idiosyncratic designs using standardized component parts, such as in the façade for the Ginger and Fred office building. It was also used to produce buildings made of discrete, bespoke parts, such as the steelwork structure in the Guggenheim Museum. Figure 6.9 illustrates the process in designing and producing the Bilbao Guggenheim Museum. In this case, designs from the CATIA software programme were automatically translated into geometric databases and digitally transferred from Santa Monica in California to the steel fabricators' BOCAD computer-numerically-controlled software

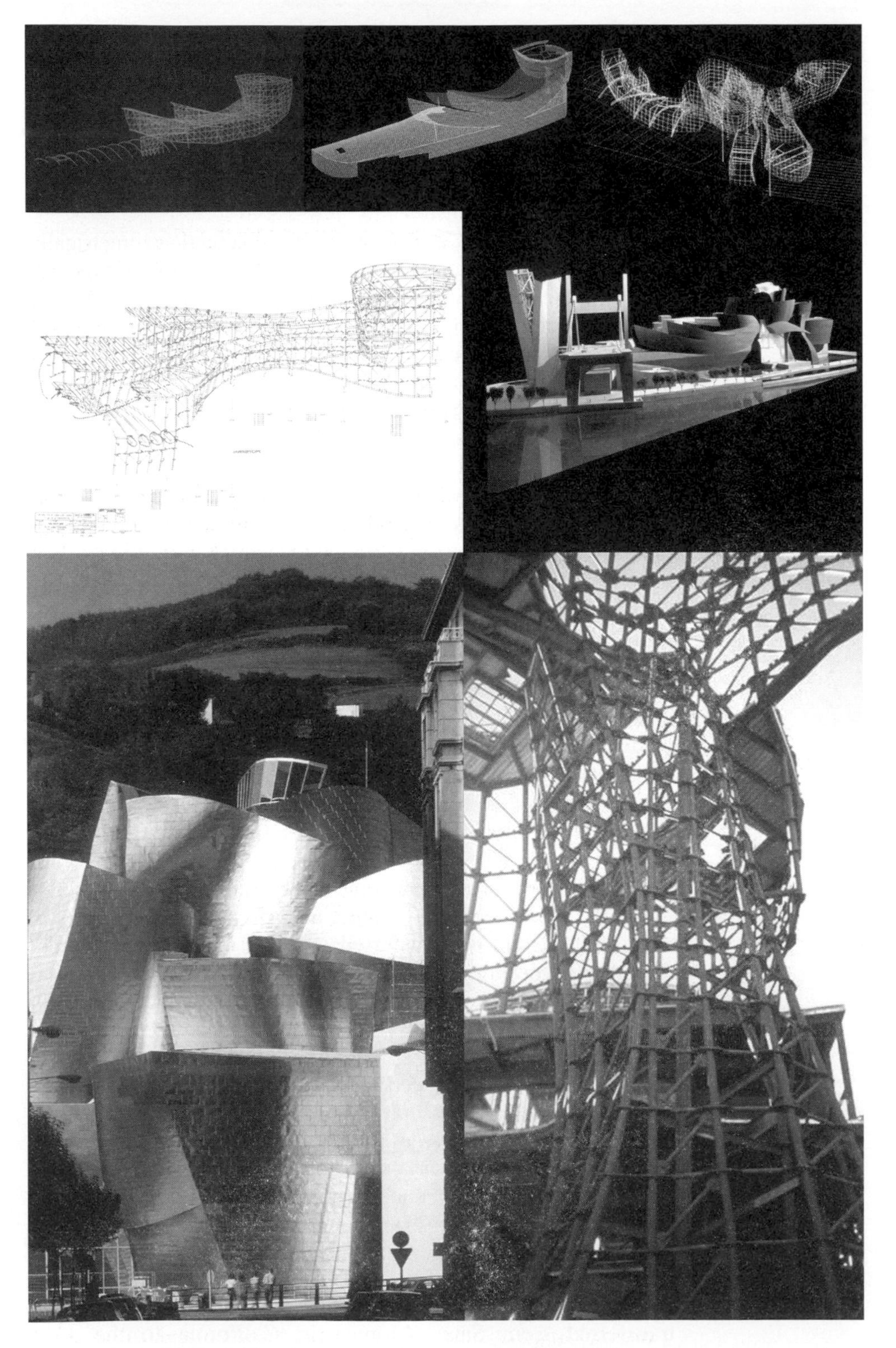

Fig. 6.9. New design and production processes: Frank Gehry's Guggenheim Museum

in northern Spain. The approach eliminated traditional steps in the process such as developing working drawings and shop drawings. Each unique piece of steel was barcoded and erected on-site using a laser surveying system linked to the CATIA model for positioning. The use of tape measures on site was unnecessary.

The starting point for this process was not the existing, traditional approach to architecture and construction. It required a radically new approach. The Gehry example combined physical and computer modelling to improve production processes in the design and construction of complex building forms. The system assisted in providing predictability from the use of repetition in manufacturing, without sacrificing form in overall design. It showed how shared databases could be used as an essential tool in new integrated design and production processes, by different specialists located in different countries. The result was a radical change in design and construction, which placed the designer within the team at the centre of the process, enabling better value to be provided to clients. The designers' responsibilities were increased because they generated the database to which everyone else worked. At the same time, the design office was placed in closer relationships with contractors, suppliers and subcontractors. This approach blurred the traditional distinctions between where architecture and engineering stopped and construction started: tasks were divided differently from those in the traditional design–construction process. Finally, the process enabled better measurement of performance improvements in terms of the number of change orders and defects in manufacturing and on-site assembly, reducing the need for cutting, adjustment and defect rectification on-site. However, in its formative stages, the development of this radically new approach clearly required the patronage of innovative clients who were prepared to underwrite the risks of innovation, such as those involved in Walt Disney's Los Angeles Concert Hall.

The Gehry example showed one possible way in which architectural design might develop in future, to become re-integrated within the total design and construction process. But this was by no means the whole story. By the 1990s it had become evident that new design skills would be needed to produce many of the complex buildings needed in a modern economy. These skills did not just encompass architecture and structural engineering, but also systems design and space planning. Previously, most architects, engineers and designers had been primarily concerned with structures, materials and systems of buildings. Whilst there were exceptions to this (for example, housing designers had attempted to deal with the use of buildings from a social and organizational perspective) in the main, designers had been primarily interested in physical properties and how they were assembled within given budgetary constraints. However, during the 1980s and 1990s a growing number of designers became interested in the design of space in relation to social uses in a variety of different markets, most notably the office sector. A number of factors resulted in a small but growing movement of architects working in new areas, such as the management of space over time. The organization of space was seen by these architects as being of equal importance to the physical development of space itself. This had brought with it a new understanding about the use of design, not merely to provide space for what organizations think they do at the moment, but for a much more active involvement in design as an agent of organizational change to improve performance in users' business processes. These developments were closely linked to wider changes in organizations and pressures to re-engineer business processes in order to achieve productivity and other performance improvements associated with the introduction of ICTs. Frank Duffy and John Worthington of DEGW were among the leaders in developing this new knowledge of the use of space in relation to organizational change.

6.4. How successful was construction in the digital age?

So far this Chapter has described technical and organizational changes in construction processes occurring from 1970 to the late 1990s. The world of construction was changing from the machine to the digital age. A range of new approaches to organizing production could be observed in different countries and these varied according to markets, technologies and specialist skills, as shown schematically in Fig. 6.10. There was no single best approach. Capabilities differed in firms from North America, Europe and Japan. During this time annual construction productivity, as measured by gross output per head, was growing at around 2 per cent in several OECD countries although the heterogeneous mix of output, cyclical fluctuations in workloads and imperfect statistical information made it difficult to measure productivity accurately. For example, there were no reliable measures of net output and employment and for these reasons only the broadest comparisons of productivity levels and movements over time could be made. Construction labour productivity growth across OECD countries between 1970 and 1993 was shown in Fig. 1.5. The figure indicates productivity growth in Portugal, Japan, the UK, Belgium, France, Sweden and Finland from the mid 1980s. However, labour productivity in Norway, Austria, the USA and Australia appeared to be falling during the same period. Labour productivity growth in construction was lower than that in manufacturing industries during the same period.

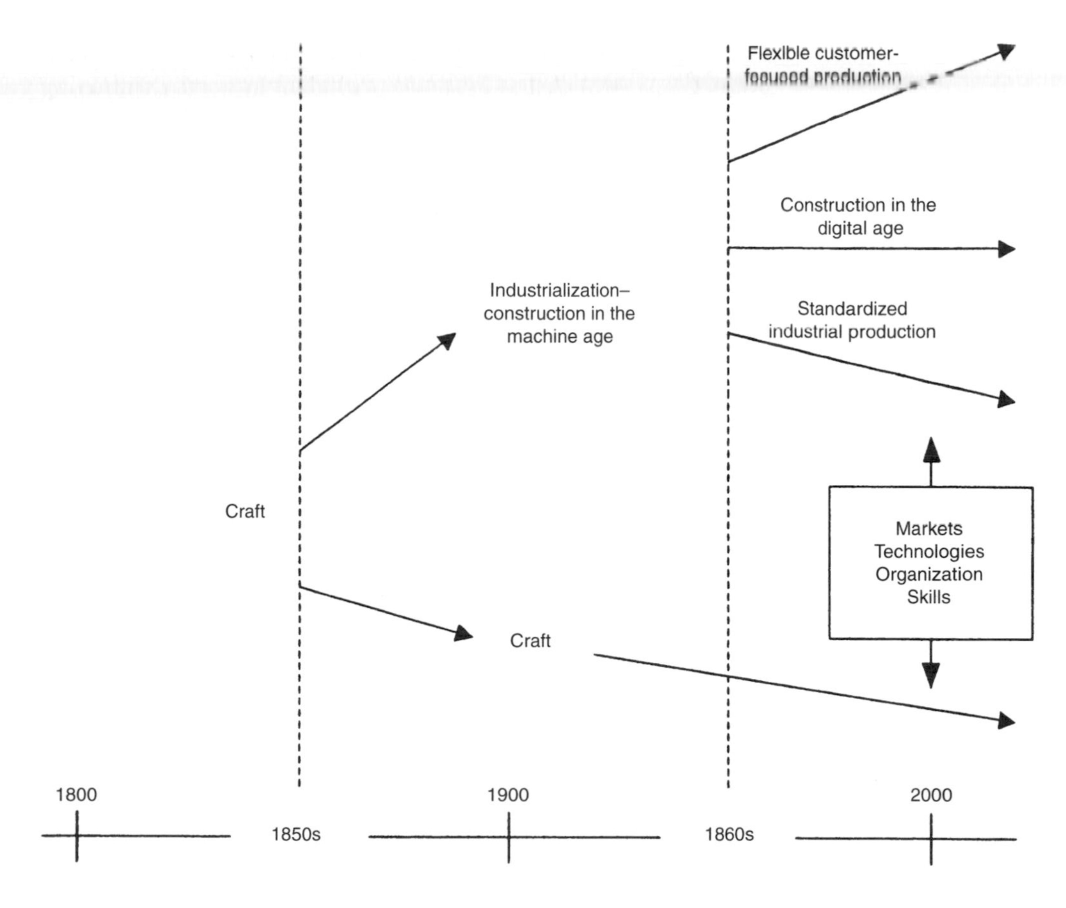

Fig. 6.10. Industrial divides in construction

Design and construction firms faced new challenges stimulated by three factors: the need to construct new types of buildings and structures; pressures from clients to improve quality, reduce costs and speed up construction processes; competition from firms introducing new approaches. The period was characterized by changes in the ways in which large projects were financed, and the proportion of private sector investment in infrastructure and building projects rose in many OECD countries.

First generation intelligent buildings constructed in the 1980s generally failed to perform to the standards required by users and expected by their designers. Problems arose because of the way in which systems were designed, installed and operated; much of the expertise was dispersed among many small firms that were poorly coordinated. Nevertheless, construction firms played a central role in design and systems integration, installation and commissioning. They functioned as conduits for technology in a process of transfer from suppliers to users. New materials and component systems were being developed and it became apparent that if complex technical systems originating upstream in the supply chain were to be used successfully within buildings it would be necessary for construction organizations to function more effectively in the transfer of technology. Changes to the design and construction system were needed in order to ensure that the assembly of integrated systems occurred properly.

A process of industrial reorganization took place in which new relationships evolved between clients, design teams, contractors and suppliers. The gap widened between firms capable of competing internationally, using ICTs, delivering technically complex projects and those producing simple buildings and structures in domestic and local markets. In order to survive, large companies based in older industrialized countries sought more work in the international marketplace as competition intensified in their home markets. Prime contractors and specialist design firms found themselves under threat from changes in procurement and project financing strategies. Yet they still needed to exercise their skills in large complex projects in order to maintain their competencies and reputations.

In response to these new conditions some international contractors changed the ways in which they operated, taking equity stakes in privately financed projects. Size was important and resulted in cross-border takeovers, mergers, acquisitions and joint ventures, particularly in European construction markets. Some large contractors were able to act as catalysts for new development work by participating in financial arrangements, inspiring greater confidence among private sector investors. Nevertheless, smaller technically specialized firms were also able to compete successfully for work in the international marketplace, in part by harnessing the benefits of new ICTs for design, specialist engineering and coordination.

The installation of information and communication systems within buildings meant that traditional mechanical and electrical contractors had to compete with component manufacturers and firms from the ICT sector to install new digital infrastructures and environmental control systems. These newcomers to construction, such as IBM, DEC and ICL, often enjoyed success because of their in-house technical expertise, but they usually struggled to comprehend the practices and conditions of construction work. They developed new products and services, sometimes on the basis of those used for their own in-house needs. During the 1980s, companies such as IBM, AT&T, NTT, DEC and Fujitsu all opened divisions or established separate firms to sell expertise gained initially through meeting their own requirements for installing information systems within intelligent buildings. For example, AT&T and NTT gained considerable experience in managing their own computer networks; at the same time they needed to ensure that sophisticated transmission and switching equipment was housed in appropriately conditioned buildings. The learning and competencies developed in meeting in-house needs provided valuable experience in the design and operation of intelligent building technologies and in facilities management.

The convergence of firms from different sectors (such as process and temperature control, mechanical and electrical engineering, electronics, telecommunications, telemetry and instrumentation, utilities and office furniture manufacturing) illustrated changes in industrial structure in the digital age. However, questions remained over how best to design and manage the installation of new systems. The resources needed for investment in new technology were beyond those that most traditional mechanical and electrical contractors could muster and they lost out to telecommunications or electronics firms in the long term. Furthermore, some traditional firms, such as electrical contractors, were at a disadvantage because they were usually too far down the chains of subcontractors to communicate effectively with clients. Specialist cabling and ICT contractors tended to have direct links with clients, side stepping the construction system, and establishing long-term relationships beyond the construction phase. Technical developments towards systems integration and intrasystem networking threatened the role of mechanical and electrical contractors further because most did not have either the resources or skills to invest in new technologies or the market power to exploit them.

There was great variation between participants involved in leading-edge methods and the rest who chose to operate using traditional craft or machine age industrialized techniques. The leading-edge operators became more international, drawing upon materials and components, labour, professional skills and management techniques from around the world. The rate and extent of technical change, demand for new specialized skills, changes in employment conditions, adoption of new management techniques, internationalization, and changes in patterns of demand indicated a sector in transition. At least three different forms of organizing work were evident: those based on traditional craft practices; those associated with the industrialized techniques used in the first machine age; and those emerging in the digital age. The differences between them are summarized in Table 6.2.

The larger, more complex and more bespoke construction activities became, the more they required coordination of inputs from a wide variety of sources. A hierarchy of project complexity could be observed, which related directly to industrial structure: as complexity increased so too did the project-

Table 6.2. Paths of industrial development from the 1970s onwards

	Craft	Machine age	Digital age
Production process	Handicraft—emphasis on cost and quality	Assembly—emphasis on cost and speed	Adaptable assembly—emphasis on quality, cost and speed
Markets	Local, indigenous markets: residential; repair and maintenance	Large and medium sized standardized projects (agricultural sheds, warehouses, off-the-shelf buildings) optimizing suppliers' costs and convenience	Mainly large projects (new sophisticated office buildings for financial services and high-technology firms). Flexible and adaptable, optimizing both clients' and suppliers' cost and convenience
Product	Bespoke, made from basic and some new materials	Standard units made from off-the-shelf prefabricated parts	Complex, made from components sourced internationally
Type of firm	Small, local, with directly employed labour	Regional or national using specialist and multi-skilled subcontractors	National and international, coordinating very specialized small firms
Design	Intuitive/vernacular	Serial process	Interactive, concurrent engineering, simulation and modelling
Materials and equipment	Local, handmade	Traditional plus technological discovery; mechanization of heavy work	Optimal choice plus purposeful design; networks of computerized coordination, flexible telecommunications and programmable tools
Competition	From other local firms	From national and international construction firms	From international construction firms and firms from other sectors, e.g. IT and electronics
Skills	Shift towards multi-skilling	Specialization of skills and de-skilling	Specialist and integration skills—new 'high-tech crafts'
Learning	Cumulative	Short formal training courses if any	Interactive

Table 6.2. Continued

	Craft	Machine age	Digital age
Innovation	Adaptation of new technologies diffusing into handicraft work	Large-scale R&D; standard responses	Large-scale R&D; flexible solutions and site-based and project-to-project learning
Technological change	Incremental adaptive change	New materials, standardization and prefabrication	Information technologies, new materials, biotechnology, pre-assembled parts
Organizational change	Erosion of traditional demarcation boundaries	Standard contracts and relationships	Experimentation with different forms of contracts and new relationships
Trajectory	Continuation along adaptive path	Deepening of existing path of industrialization	Emergence of alternative paths

based, technically specialized nature of production activities. In these markets, firms built their businesses on the provision of specialized management and technical skills, often trading on expertise and reputations accumulated over many years.

However, learning processes often remained informal with many breaks in feedback up and down supply chains. Uncertainty remained high in most building processes, because of the ways in which they were organized with weak feedback loops and poor opportunities to learn from project-to-project. The need to simplify processes, using modular component parts and standard interconnections was one response to growth in project complexity. But sometimes, simplification had its limits in terms of aesthetic appeal and functional qualities, as Robert Venturi pointed out, blatant simplification led to bland architecture: less became a bore (Venturi 1977, p. 17).

The use of modular systems was one way of managing complexity and modular components and sub-assemblies were successfully used in many forms of construction. The limits to their application related to the size and stability of the market, cost of transportation and ability to control and sub-divide labour on dispersed sites where final assembly took place. These limits were often reached earlier in highly bespoke or customized projects in which complex technical issues needed to be resolved by skilled engineers, managers and technicians on-site.

In Britain construction organizations were involved in a number of major organizational changes aimed at reducing the cost of construction in increasingly complex projects. New forms of contract aimed for closer participation between clients and prime contractors. As a consequence, the roles played by architects and other construction professionals such as quantity surveyors diminished. This 'organizational fix' resulted in changes in industrial structure in which new types of organizations such as design–build contractors and systems integrators came to the fore, while traditional professions such as architecture declined. The project-based nature of construction was enshrined within a legal and institutional framework that hindered the ability to plan the long-term use of resources and develop new technologies. Furthermore, most British contractors were involved in price-based competition, which involved keeping prices down through cost avoidance rather than by investing in process innovations. For example, overheads were avoided by moving to a casual employment system, rather than by employing skilled specialists directly. New methods of organizing work generally provided managers with the immediate short-term flexibility they required to respond to different client needs. They also involved a shift in responsibilities where the work of main contractors centred on buying in services and resources, and the physical work of construction and installation was carried out by specialist subcontractors. In 1998, British contractors continued to spend up to 10 per cent of turnover on tendering (Guthrie 1998). Problems were also experienced in managing the interfaces between tasks carried out by the many different specialist trades, and more was spent on legal fees than was invested in training.

Japanese contractors took a different approach by involving themselves in the process of technical change itself. This involved stronger vertical and horizontal linkages between design and construction organizations and their suppliers. Contractors competed through a mix of patronage, price and technological excellence. This resulted in continuity of work, both in terms of volume and type, which in turn led to greater stability, enabling them to invest more in new technology. Thus, whilst the Japanese economy was expanding until 1990, a virtuous circle was flourishing in which competition was driven by a race for continuous improvements in processes as well as products. This process began to flounder with the onset of economic recession in the 1990s. The Japanese model no longer appeared to offer the answers to the problems of performance improvement in construction. New models were needed to combine technical development and process improvement in dynamic markets. Partnering between customers and contractors and within supply chains, combined with integrated design and construction, favoured in some of the hybrid north European approaches emerging in the late 1990s, appeared to offer a rational and adaptable way forward.

PART 3

KNOWLEDGE FOR INNOVATION

7. Sources of innovation

Parts 1 and 2 of this book focused on the forces that shape innovation in the built environment. Two periods were described, each relating to particular, broad sweeping technological, economic and social conditions. Analysis of the history of industrialization in construction showed that products and processes developed along two distinct paths from the mid nineteenth century to the late 1960s: these were the craft and industrialized trajectories. During this period, industrialized buildings and construction processes were strongly influenced by the requirements, concepts and practices that developed in volume-production industries. This was the machine age, when emphasis was primarily on the development of new materials, components and equipment used in the production of building structures and envelopes. At the same time, mechanical and electrical systems installed within buildings increased in complexity. There was a close relationship between the discovery of new materials and their use in the construction of new types of buildings. Industrialized construction processes replaced many traditional craft techniques.

Whilst developments continued in the structures and fabric of buildings, by the 1970s new products, systems, services and processes emerged, associated with the advent of micro-electronics, the growing importance of active elements in buildings, such as mechatronic and control systems, and the need to produce facilities for the digital age. Industrialized construction technologies and processes were applied in a number of different ways and new influences came to the fore, challenging the traditional standardized volume-production approach. More emphasis was placed on customers, and clients sometimes led initiatives for innovation. Innovation was closely linked with the diffusion of ICTs into buildings and in design and construction processes. New materials and components created possibilities for constructing lighter weight, adaptable and more sustainable buildings. Some types of buildings became more complex and new methods of organizing construction processes were developed to enable rapid design, assembly and installation. During the 1980s and 1990s, attempts were made to speed up production, improve quality, reduce initial and lifecycle costs and limit the environmental impact of buildings and construction processes.

In the 1990s there was a growing interest within some design, engineering and construction firms and among government policy makers to understand the mechanisms by which innovation occurs and the factors that lead to success and failure in different circumstances. The traditional, static view of construction was increasingly seen as unhelpful to industry and government. This Chapter explores research on innovation in buildings and construction processes. It begins with a review of lessons from previous studies of buildings and construction. These produced useful knowledge about barriers to innovation and responses of construction firms to change, but they did not adequately explain the drivers for technological change. From an academic perspective, the built environment and construction have been poor relations in the mainstream literature on industrial innovation. Yet research on innovation from studies of several different sectors of the economy can provide helpful insights into how and why changes occur in the built environment. The Chapter continues by assessing the available literature on general patterns of innovation and how concepts might be applied to an understanding of innovation in the production and use of buildings and structures. It concludes with the development of an integrative, systems approach to innovation.

7.1. Learning from the past

Modern building science and formal research methods for understanding the performance of building materials and constructed products emerged after the First World War, when research activities aimed to broaden and deepen knowledge on specific aspects of building performance. This work sometimes resulted in the development of new building components and products. Developing a research and development capability in this emerging field was not easy, not least because it required multidisciplinary skills, drawing on knowledge from a wide variety of subjects and technologies. It was also necessary to find ways of realizing the benefits from R&D, by relating the results to practice. These issues remain important today. They were first recognized 80 years ago by those involved in forming the British Building Research Station (BRS), the world's first government-funded building research organization:

> *The problems of building research are not as a rule of such clear-cut character as those connected with say Chemistry, Physics or Engineering. Extraneous factors frequently obscure the vital points of the research and often complicate the application of the results in practice. . . . The application of the results of this research will be bound up with workmanship, qualities of material, nature of jointing material, frequency of support, and other important factors. . . . It is therefore important that research workers should be sufficiently acquainted with the circumstances in which particular laboratory results will be*

applied in practice and be able to translate them into terms of everyday practice in order that the results may be properly applied by the practical constructor. . . . It might perhaps be laid down that some at least of the building research staff should be well acquainted with practical conditions of the industry.

Source: Building Research Station, Founding Memorandum, Department of Scientific and Industrial Research Advisory Council, 1919 (reproduced from Nowak and Harling (1996)).

Initial work at the BRS focused mainly on materials and their applications. But after the Second World War, and particularly during the 1960s, emphasis was also placed on developing capabilities to research building and construction processes, including the obstacles to improving performance and productivity. Similar studies were carried out in other industrialized countries, where public sector building research organizations were established in the late 1940s and early 1950s.

Studies at these organizations were mainly concerned with the introduction of new product and process technologies, such as the potential for prefabrication and mechanization, materials handling and the use of lifting equipment. Work on construction processes tended to focus on the application of various forms of scientific management to site-based work, such as work activity sampling, and time and motion studies. The results of these proved useful in helping to optimize some parts of the industrialized construction process. They also helped to highlight impediments to innovation, such as the separation of design from production. Analysis of design and production planning processes was carried out from the late 1950s, together with the development of new process management methods derived from project management in the military sector. This work led to a better understanding of production economics and the need for feedback of knowledge from production to design, together with awareness of the need for modular coordination. By the 1960s, attention also turned to skill requirements in construction processes, resulting in a number of surveys of operative skills and attempts to re-configure work packages. Work by social psychologists on relationships and communications between players in the construction system was also important at this time.

From the 1950s to the 1980s, a number of large construction firms in the UK, Europe and North America invested in research on new materials, components, construction processes and equipment. This research either focused on requirements arising from specific projects, particularly in the energy and infrastructure areas, or was more generic in nature, such as the development of Wimpey's 'no fines' concrete. Some of these research laboratories functioned as testing stations, others also included capabilities for understanding different aspects of the economics of production. They also provided everyday technical support to designers, engineers and constructors working in the field.

7.1.1. Resistance, barriers and responses to change

The study of specific issues about how innovation occurs in the built environment has its origins in this period. The work of Marian Bowley and Ducio Turin, two academics working separately in Britain, was particularly formative. Bowley researched factors that hindered innovation, primarily in materials and components used in the construction of building structures and cladding (Bowley 1960). One of her central arguments was that major changes in techniques occurred at times when there were changes in the type of demand for buildings. Construction organizations resisted, reacted to, or remained passive recipients of such changes. Her work identified innovation from a materials supply-side perspective, for example, whether new building products would enable innovative firms to expand their markets, or reduce costs. Later, she studied the organization of production in the British building industry. She cited the 'remarkable lack of co-operation' between the various people involved in the design and erection of buildings, and an absence of informed rational choice by building owners about the types of products they required, as major barriers to innovation (Bowley 1966). She argued that the relations between owners, designers and builders were at the heart of the problems constraining innovation.

Bowley's research also explored the relationship between product and process innovation, showing that some building materials and component innovations directly affected the process, whilst others apparently involved no change in technique. These she called *ersatz innovations*, which merely acted as substitutes or alternatives to a material or component. As the palate of materials available to designers and constructors grew, there was likely to be an increasing number of ways of achieving a satisfactory result, by combining them in a range of different ways. But these innovations were to have little overall impact on performance either of the building or the process by which it was made. For example, she argued that the substitution of concrete blocks for bricks in cavity wall construction had little impact on how buildings were made (Bowley 1960, p. 33). Her point may have appeared correct when concrete blocks were first introduced, but the argument became more difficult to justify later on. Larger concrete blocks were used instead of bricks to speed up construction processes and small modifications to cavity wall designs, introducing insulation, significantly altered the performance characteristics of the final product. This example demonstrated that knowledge about the long-term impact on design and construction techniques was needed, together with an examination of the effect of changes in technology on the construction system as a whole.

The picture emerging from Bowley's work was one of a sluggish sector, lacking in internal dynamism, often passively receiving technologies developed in other industries. Her view of construction may have appeared accurate in contrast to the dynamic environment observed in other sectors in the 1950s and 1960s, for example, electrical engineering, chemicals, automobiles and aerospace. However, Bowley did not focus on the internal forces stimulating innovation from within construction, or question the implications of broad sweeping changes (such as new materials or information and communication technologies) that might affect performance across clusters of industries, including construction, or the economy as a whole. Bowley's view tended to support an argument, popular with economists and industrial historians, that construction was backward: it had not achieved the level of technological and organizational sophistication of other sectors, and had remained labour-intensive with low rates of productivity growth.

The results of research generally focused on sub-elements and the development of building products, which may have helped improve parts of the process during the 1950s and 1960s. However, it became evident that attempts to optimize components, sub-assemblies or sub-processes without evaluating the consequences for the system as a whole often only led to small improvements, or to failure. Such approaches in part resulted in an increase in building defects which Steven Groák attributed to a decline in 'robust technologies' (Groák 1992, pp. 47–49). The concept describes reliable, tried and tested methods that are used as a result of clearly understood technical precedents. These are relatively robust or insensitive to errors of design, manufacture, assembly or use. Groák suggests that many of these technologies had been pushed beyond their limits, their robustness eroded by technological, social and economic changes inside and outside construction. Introduction of new technologies in one part of the system could change other parts in such a way that traditional techniques might no longer be relied upon. The failure of feedback mechanisms to transmit information about such changes resulted in once reliable component parts becoming less stable, 'fragile technologies'.

7.1.2. Systems and total processes

This introduces the idea of buildings as complex product systems, in which changes to any one component part were likely to have implications on the design, construction and operation of other parts. Work by architects such as Robert Venturi, and other 'systems thinkers' in the 1960s helped to develop these ideas. At the same time there was a growing awareness of the need to consider the total production process. Ducio Turin recognized the need for a holistic approach to the production of buildings, calling for the study of the total process. He argued that analysis of performance in the building industry in Europe and the USA focused too much on changes in the nature of the product, in the functions of the professions, and in the contractual relationships between the participants. Only occasional reference was made to the structure of demand for buildings, the nature of the manufacturing processes underlying production of materials and components, or the standardization and quality of finishes (Turin 1966, 1967, 1975). Emphasizing the significance of the *building process* as distinct from inputs and outputs, Turin investigated the relationships between those involved in site work, clients, designers, manufacturers and suppliers. His work helped to establish new studies, which in part led to the development of the field of building economics. However, the analyses of both Bowley and Turin had limitations because they remained set within a framework dominated by the concept of *responses* to changes in demand or to technologies developed in other industries. They failed to develop an analysis of the dynamics of innovation, which included changes emanating from within construction itself.

Moreover, many studies of innovation in construction focused on what were seen to be the unique physical attributes of buildings. These were often used to explain why technological innovation is different in buildings and structures from other sectors. They included immobility, complexity, durability, costliness and the need for a high degree of social responsibility (Turin 1966). There were also regional and local differences in requirements brought about by climatic and geological conditions. This was reflected in the types of research carried out. For example, research on earthquakes has been of high priority in the USA (particularly in California) and in Japan, research on cold weather construction products and techniques has been important in the Nordic countries and Canada, and research on cooling buildings has dominated work in tropical climates.

It was often argued that physical characteristics limited the development of construction technologies (Nam and Tatum 1988). For example, immobility of the final product, necessitating its completion at the point of use, sets construction apart from the manufacturing sector where products can normally be finished in factories and then transported to the market. Immobility constrains activities to the extent that the economics of labour, machinery and transport of parts have to be considered in a different light to that in manufacturing. Complexity may stultify innovation because architects and designers may be reluctant to specify new materials and components unless they have a proven track record. The risk of failure—as experienced in some of the systems used in 1960s' high rise buildings in Britain—helped to perpetuate conservatism in design (McCutcheon 1975). In this case, conservatism was reinforced by designers' needs to take public safety into account; the assessment of consequences of design decisions for public risk is notoriously difficult. More recent examples from the digital age include the risk of 'building sickness syndrome' or legionnaires' disease that may result from inappropriate new designs in ventilation systems adopted by specialist contractors. Finally, the longevity of buildings and structures and the need for durable materials creates problems for testing new materials and components. It is often difficult and expensive to simulate the effects of weathering and other durability aspects on materials and components designed for a 60-year lifespan. The costs involved may render innovation prohibitively expensive. Furthermore, whilst individual components may be tested, it is more difficult to test the way in which they function together as systems in completed buildings. This is because prototypes of buildings and structures are rarely constructed and tested in the manner found in other industries such as automobiles or aerospace.

Despite these reservations, examples of innovations described in Chapters 5 and 6 illustrate that the physical characteristics of construction do not necessarily relate to the sector's presumed slow adoption of new technologies. Just as these characteristics may hinder the development of new techniques, they may also play a part in promoting change. For example, one of the reasons for the development of prefabricated, off-site production was to reduce the impact of dirty and dangerous physical conditions found on sites. Technological change and the physical characteristics of construction therefore influence one another in a more complex and systemic way than was often presumed. The risks of innovation may retard development in one direction, but they may also expose the need for change in others. While construction may have a number of unique features, it is by no means the only sector attempting to produce complex, durable, costly and safe products—ships, aircraft and medical equipment are other examples. There is therefore a danger in overemphasizing the limits to innovation caused by physical characteristics of construction, particularly when considering technological change. A systems approach, which examines the total process, including physical and social relations of production provides a richer understanding of the innovation process.

7.1.3. Research and development

During the 1970s and 1980s, construction research grew and diversified in university departments such as in civil engineering, architecture and the newer disciplines of building and construction management. Large European construction companies, such as Skanska and Bouygues, also increased their research activities, and from the early 1970s onwards, major Japanese contractors opened research facilities. Meanwhile, research in government funded laboratories changed focus. In Britain, research on construction processes declined at the Building Research Establishment (BRE) and much of its efforts were placed in testing materials and components in support of regulatory processes.

A serious lack of construction process research was identified in Britain in the early 1990s (Gann *et al.* 1992), and research programmes were put in place to encourage new forms of public and private collaboration in research. Additional emphasis was given to construction process research, particularly within universities, following the publication of the UK Government's Engineering White Paper *Realizing our Potential*, which resulted in the formation of the Innovative Manufacturing Initiative's 'Construction as a Manufacturing Process' Programme in 1994 (EPSRC 1994). The Latham and Egan Reports also focused attention on these areas in 1994 and 1998 (Latham 1994, Construction Taskforce 1998). New research themes emerged, once again involving social scientists working with managers and engineers to benchmark performance and understand business and project processes. These changes also led to new approaches to research on user requirements, 'value management', space planning and post-occupancy satisfaction surveys. Process research involved supply chain analysis, including work on time compression, logistics and the use of ICTs for coordination. Computer simulation and virtual reality tools also became important in the research portfolio. Perhaps the most important shift in research since the 1960s was from the analysis of particular individual work items to process mapping and total process analysis.

In spite of what appeared to be an obvious need for more research, by the 1980s most construction firms in the UK and USA had reduced their R&D expenditures as they attempted to cut overheads in their search for flexibility. A survey of design and construction firms in the USA during the early 1990s found that they invested 0.5 per cent of total revenue on R&D compared with between 3 and 4 per cent for a composite of major industries (Bernstein and Lemer 1996). Meanwhile, a number of continental European firms continued to invest in their research capabilities. Yet, despite a decline in private sector funding of R&D during the 1980s, most technological development continued to take place in R&D departments of industry (mainly manufacturers of construction materials and components) where new technology could be built upon existing technology. Data on the amount of funding of construction-related research indicated that private firms carried out the majority—about 60 per cent in the UK in 1994 (Gann *et al.* 1992, CFR 1996).

The location of private investments in construction R&D is significant. The majority of private sector construction R&D was carried out by materials and components producers, who generally developed products aimed at improving the performance of buildings and structures. Very little R&D towards improving construction processes appeared to be carried out by construction firms. This shortcoming was increasingly recognized by industry and government as adversely affecting the ability of construction to integrate new technologies developed upstream by manufacturers (Gann *et al.* 1992).

Figure 7.1 illustrates the level of expenditure on construction R&D in OECD countries as a proportion of construction output. Japan and the Nordic countries are notable for their higher expenditure in comparison with other countries. The UK performs badly—investment in R&D is more than three times lower than main European rivals and seventeen times lower than that of Japan.

7.1.4. Research in Japan

Investment in R&D was very different in Japan. The Japanese Building Research Institute was established in 1946. In many

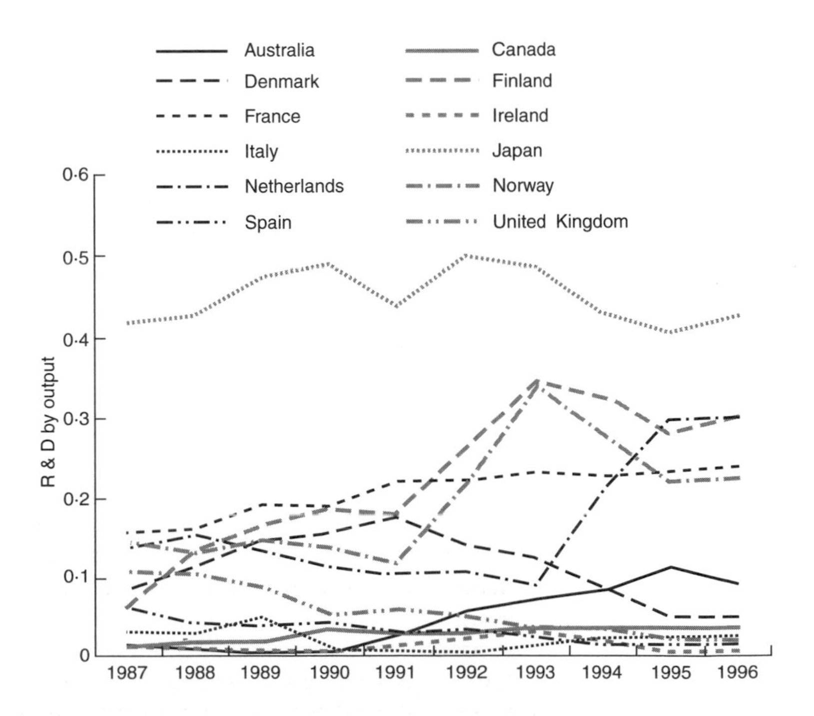

Fig. 7.1. OECD construction R&D expenditure as a proportion of construction output (source: OECD Accounts and ANBERD data. Current prices and national currency)

Note: No recent data available for Germany or USA. Approximations have been used in a few cases where annual data was not available.

respects it performed similar roles to other national building research organizations, whilst building research in the university sector remained minimal (Kose 1997). Between the 1950s and 1990, however, Japan experienced a period of urban development and infrastructural modernization on a huge scale that resulted in large volumes of work for construction. Innovation in the Japanese construction system illustrated a different approach to the development and use of technology, in which contractors, design organizations and specialist subcontractors initiated change. Changes in Japanese construction products and processes were driven by technological competition between firms, and not solely by price-based competition as tended to be the case in the UK and the USA. Firms from a diverse range of Japanese industries including real estate companies, large construction corporations, specialist subcontractors and component suppliers, banks, electronics and telecommunications firms and office equipment manufacturers, contributed to the development of intelligent building technologies. This resulted in a technological race to develop 'show case' buildings that displayed firms' technical expertise in design, construction and systems integration.

The emphasis placed on the development of new technology by Japanese construction organizations was matched by their commitment to investing in R&D. In 1989, the ratio of R&D expenditure to value added by Japanese construction was approximately 1 per cent, compared with a ratio of less than 0.1 per cent in the UK. In the year ending March 1993, Kajima Corporation, one of the largest contractors, spent over £125 million on R&D. By comparison, *total* R&D expenditure of all UK large contractors in 1992 was around £17.5 million. New construction process innovations based on the results of in-house R&D by construction firms themselves, rather than their suppliers, began to emerge. Privately owned laboratories carried out research for product testing and standards compliance, research into construction processes, including robotics, work on business processes and work organization, and research into customer preferences and user needs. From the late 1980s onwards, large Japanese housing manufacturers/builders established comprehensive R&D institutes with capabilities ranging across the physical and materials sciences and the social sciences, including economics, sociology, psychology and ergonomics. Japanese contractors had developed close and long-term links with customers and end-users and they generally had better links between marketing and production and were more involved in process innovations than their western counterparts (Westney 1987).

The practices of Japan's specialist subcontractors also differed in a number of important respects from their western counterparts, increasing their incentives and opportunities to innovate. One of the main differences was a closer association between specialist contractors and their clients. For example, in the early 1990s, eight of the ten largest electrical contractors were affiliated with the ten electric power generation companies operating regionally across Japan. The next tier of electrical contractors was affiliated to the railway companies and the third tier of smaller firms was affiliated to major clients such as highways or steel companies. Smaller subcontractors usually had close links with one or two larger specialist and/or main contractors. Close links with clients helped to provide stable revenue streams in known areas of work and the development of technology was often focused on meeting clients' particular needs. Collaboration in R&D, working in joint venture organizations and an attitude of cooperation rather than conflict was particularly important. Collaboration in R&D meant that there was often greater vertical transfer of technologies up and down the supply chain than was found in the UK or USA. This also resulted in the development and transfer of technology between manufacturers, contractors, clients and end-users occurring more rapidly and without some of the problems caused by separation of different specialist areas, found in the West. Joint venturing also resulted in better horizontal transfer of technologies between one specialist contractor and another. Unfortunately, recession in the 1990s resulted in a decline in investment in R&D and the potential demise of integrated, interdisciplinary research capabilities.

7.1.5. Research in the 1990s

In the UK, government policies towards building and construction research began to change. The role of research and development as an input to processes of technological and organizational innovation was increasing in importance during the digital age, in tandem with growth in technical and operational complexity of buildings. However, increased importance did not necessarily result in increased investment. In 1990, the BRE's relationship with the British government changed when it became an Executive Agency, which meant that it could operate as a self-financing business, earning revenue from non-governmental sources (Courtney 1997). The BRE was privatized in 1997, following a trend in which TNO (The Netherlands), VTT (Finland) and BRANZ (New Zealand) had already been aligned more closely with industry, stimulating further debate about privatization and/or commercialization of national building and construction research capabilities (Lorch 1997, Seaden 1997). Traditional public sector research institutions were therefore undergoing considerable changes. Previously, some research laboratories had been able to add value because of their neutrality and impartiality as advisors. Others had created an environment for learning and the location for joint ventures between firms and researchers, which had proved fruitful because government paid for some of the risks in research. But new models of privatized non-governmental laboratories were emerging and there was a great deal of uncertainty over how they would perform. These institutions had to grapple with questions about independence and vested interests, including the legitimacy of advice and the efficacy of new research practices.

Across Europe, general trends were becoming evident towards convergence in areas of public sector research, including a blurring of boundaries between the roles of university research and long-term, mission-oriented research carried out by institutes and government research laboratories which constituted centres of expertise in specific areas (Senker 1998). These changes raised concerns about the development and funding of long-term research capabilities. Whilst research was regarded by policy makers and funders as an input to wider processes of innovation, little research was being carried out on issues such as how to manage innovation within the business processes of project-based firms—despite the major changes faced by many design and construction companies. Research on the mechanisms and policies for construction innovation remained comparatively weak.

In the late 1990s, difficulties remained in answering questions about some of the reasons why and for whom building and construction research was being carried out. The scope of research activities and knowledge about the means by which research capabilities should be organized remained uncertain, as did knowledge about the best means of transferring results into practice. The works of Bowley and Turin provided useful conceptual frameworks for understanding these issues, but they also appeared increasingly out-of-date. They needed to be supplemented with a better understanding of technological and economic changes in the context of new social and institutional structures.

7.2. Technological trajectories and lessons from different sectors

In Chapters 2 and 4 it was argued that generic technologies, which resulted in many changes throughout the economy and society, also had a profound effect on construction and the built environment. The transition from the machine age to the digital age corresponded with requirements for new types of buildings, new production technologies and different ways of organizing processes. But how might such changes be explained? Why does the rate of change vary between and within different parts of construction and through the supply industries? How have some processes and products from different eras seemingly continued to exist through successive epochs, as shown in Fig. 6.10?

Freeman and Soete (1997, pp. 65–70) argue that each successive period in the history of industrialization relates to major new technological capabilities. For example, between the 1830s and 1880s, the development of steam engines and railways resulted in major transformations throughout society and the economy. Similar consequences could be observed with electrical and heavy engineering from the 1880s to the 1940s; petrochemicals, oil, automobiles and transportation from the 1940s to 1980s; and ICTs from around the 1970s and 1980s onwards. Each period required new types of buildings and structures to provide the facilities supporting the economic activities based on these technologies. Chapter 4 described the technological and economic changes associated with ICTs. Freeman (1991, p. 25) describes these major shifts in underpinning technologies which have such pervasive effects throughout the economy, as constituting a change in 'techno-economic paradigm'. This concept is used by a growing number of evolutionary economists and economic historians to explain radical transformations in prevailing engineering and managerial practices, applicable to almost any industry. The concept refers to changes in the body of knowledge and practices associated with established patterns of production and consumption, together with the emergence of new 'best practice' sets of rules and customs for designers, engineers, entrepreneurs and managers.

Freeman argues that such macro-level paradigm shifts embody the evolution of several 'new technology systems' (constellations of technically and economically interrelated innovations affecting branches of the economy), which themselves may involve many clusters of radical and incremental innovations. But for such a techno-economic transformation to occur, a key factor is necessary which simultaneously fulfils the following four conditions. First, it should have clearly perceived low and rapidly falling costs. Second, it should be in almost unlimited supply over long periods. Third, it should have the potential for use and incorporation in many different products and processes. Fourth, it should offer the ability to reduce costs and change the quality of products and processes (Perez 1983, p. 361, Perez 1985, p. 444).

The key factor of low-cost energy in the form of oil gave rise to the energy-intensive, volume and flow production paradigms of the 1950s and 1960s. ICTs emerged in the 1960s and began to penetrate most industries and services by the 1970s, heralding the change in paradigm to an information-intensive, flexible, computerized technology characteristic of the 1980s and widely used in the 1990s. The key factors in this new paradigm were low-cost microelectronics, information-intensive activities and associated new value-added services. The effect of introducing microelectronics was comparable to the electrification of industry at the beginning of the century, in that electric power had the capacity to transform other existing sectors of the economy as well as to impel the growth of entirely new industries (Freeman and Soete 1988, p. 2). In future a new techno-economic paradigm may develop as a result of innovation in the biological and molecular engineering fields— the biomolecular age, discussed briefly in Section 8.4.

If a new techno-economic paradigm offers such significant cost and performance advantages over existing systems, the question remains as to why diffusion does not occur almost immediately throughout the economy and across sectors like construction. The concepts of 'mismatches' between new technical opportunities and existing institutional structures, together with 'bottlenecks' in technological developments, are used to explain some of these constraints to rapid diffusion. Nathan Rosenberg (1976, pp. 125, 201) uses the concept of bottlenecks to analyse the reasons why firms carry out innovations to overcome the constraints that restrict their operations. He argues that as new production activities expand, they run into a series of problems which must be overcome before further benefits can be accrued. As one bottleneck is overcome, another eventually asserts itself. Efforts are then made to alleviate these problems, often leading to the use of innovative techniques. Thomas Hughes uses a slightly different approach to explaining opportunities and barriers, based on the military metaphor of reverse salients, described in Section 3.4.

7.2.1. Social and institutional inertia and rates of diffusion

These conceptual models of broad sweeping innovation processes have proved useful in government and industry, helping policy makers focus their strategies in support of innovation within a historical framework which considers technical, economic and institutional factors in different sectors. They also appear useful in helping to explain long-term changes in construction and the built environment. However, the relationships between new technical and economic opportunities and the social, institutional and organizational conditions required for exploiting them needs further explanation. For example, why is it so difficult to introduce seemingly useful information systems across a team of building designers and engineers?

Carlota Perez argues that, in order to fulfil the potential of a new techno-economic paradigm, a fundamental transformation of existing social and institutional structures is normally needed at national and international levels. She defines technology as 'the how and what of production', and points out that this is partly a social and economic matter, determined by prevailing social and organizational conditions. Yet, because existing 'normal' patterns of social behaviour and institutional frameworks are shaped by the requirements of accommodating the most widely used technologies, mismatches occur when the potential of a new technology emerges. There is often relative inertia to its use in existing firms, organizations and institutions. The failure, or inability, of social and institutional changes to keep pace with techno-economic possibilities is an obstacle to the full deployment of potential benefits. For example, the implementation of major new technological systems usually necessitates changes in firms' organizational structures and the relationships between firms, as well as the regulatory environment within which they operate. The use of CATIA design and engineering tools, together with Internet technologies, by Frank O. Gehry Associates (described in Section 6.3) illustrates this point.

Established interest groups, such as professional institutions and industry associations, are also often challenged by such developments. The time needed for institutional and social readjustments, skills training and organizational developments to take place helps to explain why new techno-economic paradigms rarely emerge. When they do, this in part accounts for why it takes a long time for the full social and economic effects to be realized.

However, when firms and organizations are able to prove substantial benefits from new technologies, others quickly strive to follow suit. The diffusion of major new technologies happens through a process of 'swarming' in which imitators begin to realize the potential for profitability of new products and processes. A bandwagon effect takes place as more firms become involved. The process of imitation is not that of simple replication, rather, it involves further major innovations and smaller refinements. Once a new technological regime is established, a string of further innovations follows along the new trajectory. Clusters of complementary innovations develop resulting in new technology systems whose characteristics pertain to different uses in different sectors (Freeman *et al.* 1982, pp. 64–81).

The rate and extent of diffusion has been shown to be uneven, partly because older industries, like construction, lag behind modern advancing sectors. Older industries often take longer to make the necessary organizational and technical adjustments and overcome bottlenecks to the application of new systems. For example, traditional management styles and market structures geared to the machine age have hindered the adoption of ICTs in construction. The rate of diffusion from leading players to others is therefore heavily dependent on the ability of firms and industry as a whole to make the necessary organizational and institutional changes.

Nevertheless, institutional structures should not be regarded purely as barriers to change. Firms and institutions play a role in reducing uncertainty by providing sets of routines and rules that regulate the relations between organizations. There are limits to how quickly they can change without disrupting society but, at the same time, they provide the stability necessary for technological change to occur along established routinized paths. Even radical technological change depends on institutionalized behaviour patterns. So institutions provide both positive and negative stimuli to innovative activities (Johnson 1992, pp. 23–29), and institutional and organizational innovations may precede technological changes in some cases. For example, in the 1990s, many efforts were being made in the UK and the USA to change institutional and organizational processes in construction (such as through partnering) as a prerequisite for the introduction of new technologies. In Britain, the Construction Taskforce report published in 1998 focused primarily on improving processes—the idea being that once better ways of organizing construction had been established it should be easier to introduce new technologies.

7.2.2. Trajectories and sector studies

This framework helps to explain the nature of technological and economic changes brought about by the introduction and diffusion of major innovations such as ICTs. It shows that existing social and institutional structures constrain the rate of diffusion, particularly in older sectors such as construction. But more detailed work is needed to understand the mechanisms by which changes occur within a particular sector. A general model of change associated with progress along technological paths was developed in a study of the semiconductor industry (Dosi 1982). Giovanni Dosi began his work by questioning the validity of demand-pull and supply, or technology-push theories of innovation. Neither theory accounted adequately for the complex structures of feedback between the economic environment and the directions of technological change. There was no linear process of development from research to application. Whilst technology-push factors may be important in the early stages of development in some technology-driven sectors, these appear to be progressively moderated by demand pressures over time (Walsh 1984, Rosenberg 1991). In searching for a more comprehensive picture of technological change, Dosi used the concept of 'technological trajectories', defined as progress along paths based on specific technological and economic trade-offs. Once a paradigm became established it had a powerful 'exclusion effect' because of the narrow focus of activities that became routinized as accepted wisdom, thus reducing the search for alternatives. Progress therefore tended to become path dependent.

The term 'technological trajectory' describes the direction of advance within a technological paradigm. This is derived from the patterns of normal problem solving activities found once an industry has become attuned to using particular technologies. The concept of 'selection devices' is used to explain the specific direction of technological development, and the reasons why some solutions are selected and others rejected. These include social, institutional and economic factors, such as cost-saving capabilities, which are often found to be stronger than market mechanisms at the selection stage of a new technology. They may also be negative events, such as disasters caused by technical failures that spur further innovations in their resolution. Using a biological analogy, Dosi viewed changes of technological trajectory as similar to mutation in the natural environment. He suggested that social and economic environments affect the emergence of new technological trajectories first by selecting the direction of mutation and then by selecting among different mutations (Dosi 1982, pp. 155–156). Once the new direction of development has gained momentum it proceeds through processes of technological and organizational adaptation. Using these concepts, it is possible to identify two phases in the development of an industry. First, there are the early stages of the emergence of a new trajectory, typified by economic 'trial and error' in which new firms emerge, driven by the search for new profits. This was evident in the early stages in the development of intelligent building technologies. Second, in more mature stages of development, oligopolistic structures emerge, in which technical change tends to become endogenous to firms' competitive behaviour, based upon the search for new technological advantages along a more defined path (Dosi 1982, p. 157). This process was evident in subsequent stages in the development of intelligent buildings, where, for example, controls manufacturers had been able to secure particular market advantages and positions.

Three characteristics of innovative environments emerge from this work.

- Innovation is an *evolutionary process*, developing along particular paths, shaped by social and institutional conditions. Evolution occurs along a particular path determined by economic trade-offs.
- Innovation is *an irreversible process*. Once a path has become established it is not possible — for economic, technical and organizational reasons — to move back to former practices. For example, engineers no longer use slide rules to make their calculations.
- Innovative environments are *self-organizing* in that the order of the system is largely unintentional, emerging through the dynamic interactions between technological progress, economic activities and institutions governing decisions and expectations.

This framework is helpful in analysis of discontinuous changes in a techno-economic paradigm and for explaining incremental, continuous technical progress along particular paths. It helps to account for radical and more incremental changes brought about by the development and introduction of digital control systems in buildings, or information and communication systems within construction processes.

7.2.3. Technological selection in construction

The development of semiconductors was the focal point of Dosi's study. Technologies in this sector resulted from intensive R&D processes, before products were exposed to the types of social and institutional selection which affect many other innovations at an earlier stage in their development. The use of semiconductors resulted in radical new innovations leading to a new techno-economic paradigm, based on microelectronics, computing and telecommunications. The study proved useful in explaining the evolutionary nature of industrial development and the selection environments within which choices of technologies are made. Most of the sectors to which these theories have been applied in subsequent studies invest substantially in R&D, and have generally well developed competencies in the management of technology both within core firms and in closely related supply firms. Furthermore, the selection environment in some of these leading industries may be more clear-cut than in others such as construction.

By contrast, construction firms spend little on R&D and technical competencies are not usually organized in such a coherent manner to exploit the benefits of new technologies. As a result, selection environments in construction have their own particular type of complexity, which often leads to inefficient methods of choosing technology. Selection in construction involves different levels of bureaucracy, such as the need for new technologies to comply with building regulations developed to govern the use of older technologies. Furthermore, returns on investment in new techniques, in the form of increased profitability, may not be immediately obvious. Returns on investment to users may take years to pay back, as is the case with some energy-saving technologies.

Whilst studies by Freeman, Perez, Dosi and others show that the ability to develop new technologies has been of strategic importance to firms' competitiveness in industries which are research-intensive, far less is known about the role of technology in sectors like construction where formal research is generally of low priority. The difference between firms' abilities to assimilate benefits from new technologies is one of the reasons why diffusion occurs more quickly in some sectors than in others. The diffusion of microelectronics appears to be easier in more technologically homogeneous sectors such as telecommunications. Such patterns of change are more difficult to understand in the production and use of the built environment, where it is necessary to differentiate between those activities where there are capabilities in managing processes of technological change and those without such competencies.

7.2.4. Knowledge requirements and patterns of innovation

Production processes can be categorized into five different generic organizational forms—project, job shop, batch, assembly line and continuous flow processes (Woodward 1965, Hayes and Wheelwright 1984, pp. 176–183). Construction activities are examples of the project-based organization of production, which requires the coordinated input of a wide variety of resources to produce customized products. Industrial structure is dictated to a large extent by the project-based nature of the activity. Firms build their businesses on the provision of specialized management skills and resources that often rely upon expertise accumulated over many years. However, learning processes are usually informal with many breaks and little feedback upstream, downstream or to other construction organizations. Capital intensity is generally low. Professionalization, codes of practice, standard procedures and building regulations, together with traditional craft demarcation lines upheld by trades unions and employers create a 'locked system' which is slow to change (Nam and Tatum 1988, p. 140). Tasks often vary from project to project and site environments do not correspond to those found in factories: the need to combat adverse weather conditions and the seasonal nature of work are often cited as hindering the development of construction techniques. In contrast, job shop forms of organization are used by firms producing small batches (and often a large variety) of products, requiring different processing sequences, for example, machine tools. Batch production methods are used to produce more standardized products such as metal castings, while assembly line processes are used in the production of standardized products such as automobiles, digital watches and many electronic components. The organization of production in continuous flow processes is used in high-volume chemical plants, oil refineries and food processing; this is very capital-intensive requiring high equipment utilization rates.

Patterns of technological change vary depending upon the nature of the production systems and different sources of knowledge, which usually relate to the size and activities of innovating firms. For example, some firms specialize in particular technical fields and have strong capabilities to innovate in these areas, such as some of the specialists supplying technologies for intelligent building systems. Others may operate in a multi-technology environment or manage systems integration; in this case they need technical capabilities in a range of areas. Models of technical change focusing on firms' competencies are therefore needed to understand how they might manage their resources (Teece and Pisano 1994, Tidd *et al.* 1997, Pavitt 1999). When firms innovate, they often have to extend existing capabilities and build new competencies. For this reason, adapting firms' capabilities to reap the benefits of new technological opportunities usually takes time. The process is often a cumulative one, based on the extension of existing skills and acquisition of new capabilities. Keith Pavitt argues that what firms can realistically try to do technically in the future is strongly conditioned by what they have been able to do technically in the past. Observing a number of industries, Pavitt developed a taxonomy of different types of firms, defined according to whether they are supplier-dominated, production-intensive, or science-based in relation to their capabilities to develop and use new technologies (Pavitt 1984).

This taxonomy is useful in that it begins to articulate linkages in the transfer of technology between sectors. For example, according to Pavitt's taxonomy, construction firms fall into the supplier-dominated category. They often receive their technological inputs from science-based and scale-intensive firms (e.g. chemicals, materials and component manufacturers). The advent of customer-led innovation for intelligent buildings also shows that the nature of innovation in these businesses is demand-induced. To operate successfully and benefit from new technological opportunities, design and construction firms therefore require the capability to access and adopt technologies developed upstream in the supply chain. They also require strong market- and customer-focused capabilities to ensure that developments are appropriate in meeting users' requirements, and that lessons from use are fed back to those developing the next generation of technologies.

However, there is an issue about how technology flows between different sectors are measured and managed. In much of the literature on innovation, metrics include R&D expenditure, patenting activity and bibliometrics. But as we have seen, design and construction organizations do not invest much in R&D—construction is a process which, in the main, transforms the technologies developed elsewhere. Many firms' main activities involve knowledge processes of design, engineering, coordination and integration of technologies developed in the supply industries. Previous Chapters have illustrated that some construction firms are particularly innovative, but traditional quantitative indicators do not take adequate account of these sources of innovation such as information generated in design and engineering or in pre-project surveys and feasibility studies, found in project-based activities. Neither do they help in understanding innovation in the development of new services associated with the management and use of complex products.

Moreover, small adaptive changes are of significant importance in project-based activities. Without them, many projects could not be completed successfully. In her study of stressed skin housing panels, Sarah Slaughter found that what she calls 'user-builders' made many low-cost, unique solutions to problems arising on-site (Slaughter 1993). Builders were the exclusive source of innovations relating to the integration of components. Furthermore, manufacturers had failed to commercialize these innovations. Such innovations occur on-the-job, often as a result of routine problem solving activities, sometimes by firms acting together and across different disciplines and sub-systems. The cumulative impact of these small changes may lead to improvements in the performance of a system, but on their own they are generally of limited significance (Rosenberg 1982, Ch. 3). These more informal sources of change play a significant role in the development of new technologies and methods within construction. However, they are difficult to define, observe and measure. The infrastructure in which development work takes place is less visible to conventional surveys because it is not documented in a structured manner. Project-based firms could benefit by developing indicators that would allow them to measure and manage these types of innovations within the context of a more strategic approach to major techno-economic and socio-institutional changes.

7.3. An integrative, systems approach

Part 2 of this book highlighted four significant factors in the evolution of intelligent building technologies. First, many technologies were developed initially by manufacturers and suppliers, some of whom had little experience in construction markets but whose search for new market opportunities prompted them to consider developing construction products. Second, intelligent building technologies only became effective after a period of adaptation and development by other organizations in the supply chain. This often involved closing the gap between project-based approaches to on-site installation activities and the factory-based design, manufacture and assembly of expert systems. This in turn had consequences for many construction activities involving fixing components in assembly-line processes rather than working with basic materials in traditional craft tasks. Third, successful outcomes usually involved close interaction between users and producers. Indeed, clients and users played an important role in influencing the direction of technical developments once they had realized the advantages to be gained from new functions offered by intelligent building systems and from economies in the way in which these could be operated. Sophisticated users were seen to play a significant role in the development of new systems, by specifying their requirements and working closely with organizations in supply networks. Fourth, successful implementation of intelligent building technologies was usually only possible when design and construction organizations, sometimes operating in pivotal positions as systems-integrators and installers, changed their methods and became involved in innovative activities themselves: this usually resulted in new organizational forms. In some countries, most notably Japan, government also played an important role in creating an environment within which such developments could take place.

Within this integrated systems approach to innovation in construction, the position of players in the supply chain, their relationships with users and methods by which they derive profits all affect their incentives to innovate. Figure 7.2 illustrates the main players in the British construction system and the relationship between profit sources and incentives to innovate. The arrow indicates flows of products and services from upstream manufacturers in the supply network to end-users of buildings.

The level of engagement in R&D by the majority of construction and design firms at best amounted to changes in their processes, acquisition of new skills and the development of capabilities to manage new technologies. For example, requirements to manage more complex projects in the digital age led a number of general contractors to change their market positions by developing and offering different services. They shifted away from direct responsibilities for construction work and instead provided business services in a variety of forms of construction and project management. Many other firms either ignored or resisted change and they were to face increasing competition from those who took more innovative approaches.

Our understanding of innovation in construction processes is far from complete. Five elements of the above analysis require more detailed attention in future studies and these are now described.

(1) The relationships between suppliers of components and equipment, those involved in design and those responsible for construction are not clearly enough articulated with respect to understanding mechanisms of technical change and transfer in construction. This is important particularly where buildings are constructed from many hundreds of sub-systems which themselves are made of thousands of component parts and it is not always clear at what stage in the production process innovations leading to successful outcomes are made.

Construction is the integration and assembly phase in a process involving flows of products and services from upstream extractive and manufacturing sectors through to their final point of consumption by end-users of the built environment. There are often long supply networks in

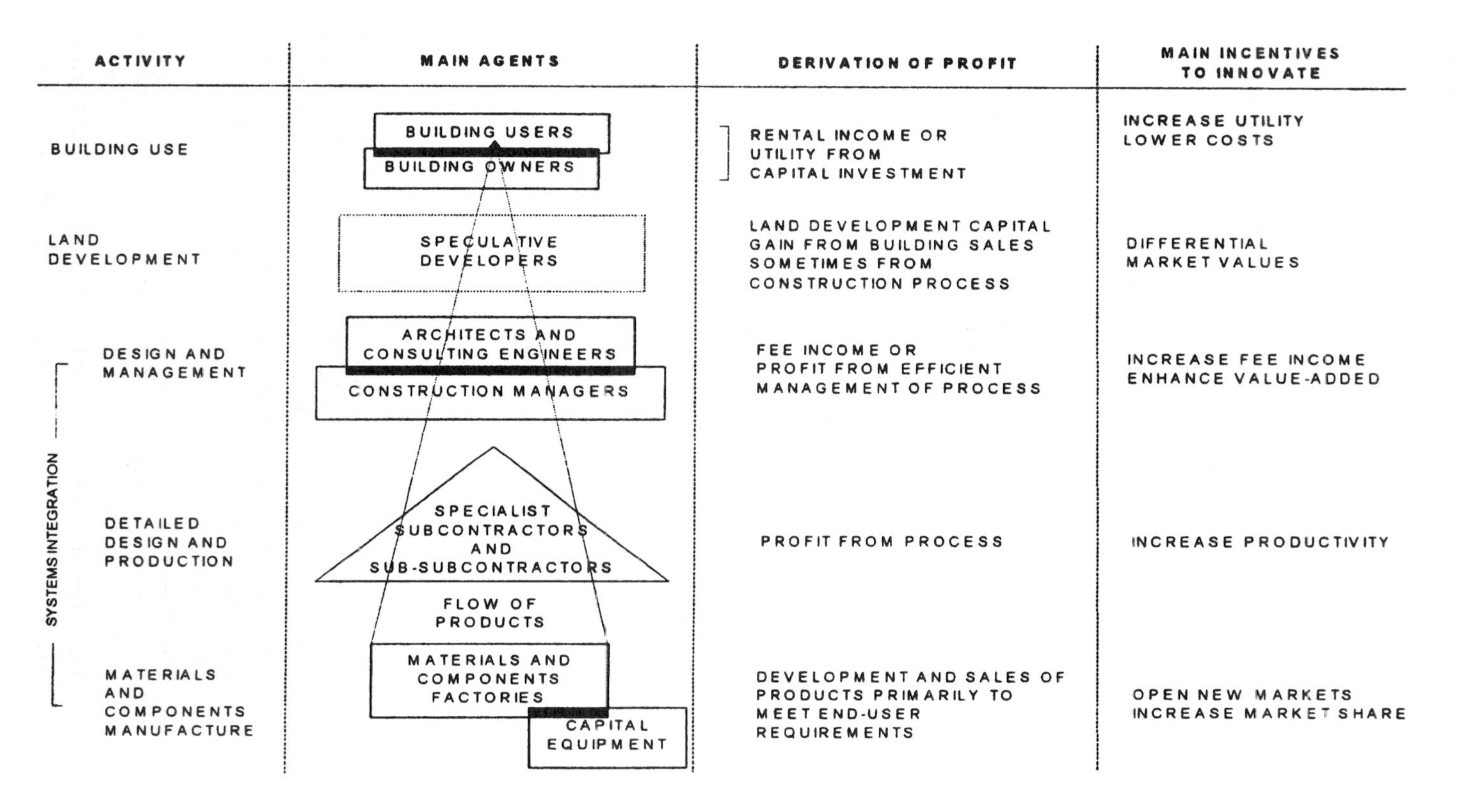

Fig. 7.2. Incentives to innovate

which much of the formal R&D activity is carried out. Technological interdependencies and complementarities exist between clusters of components that are combined during sub-assembly, construction and installation. As a consequence, the *flows of technology* within the construction system exhibit a more complex pattern than the *commodity flows* that result in the final products (Fig. 7.3). Construction processes involve systems integration that includes inter-sectoral flows of technology: when technical change occurs, boundaries between old and new construction activities and other industries shift in a process that often results in the combination of existing technologies in new ways. The concept of 'technology fusion' helps to explain the type of systems integration occurring in construction (Kodama 1992). Firms make use of existing technologies that come from different industries, they may combine them in such a way as to provide significant improvements in products or processes.

Construction firms therefore relate to many other industries in the supply stream, together with end-users and with government and industry organizations through particular technology and information flows. But whilst many technological transactions occur in construction activities, the networks of relationships between different players are not always explicit. In order to analyse these

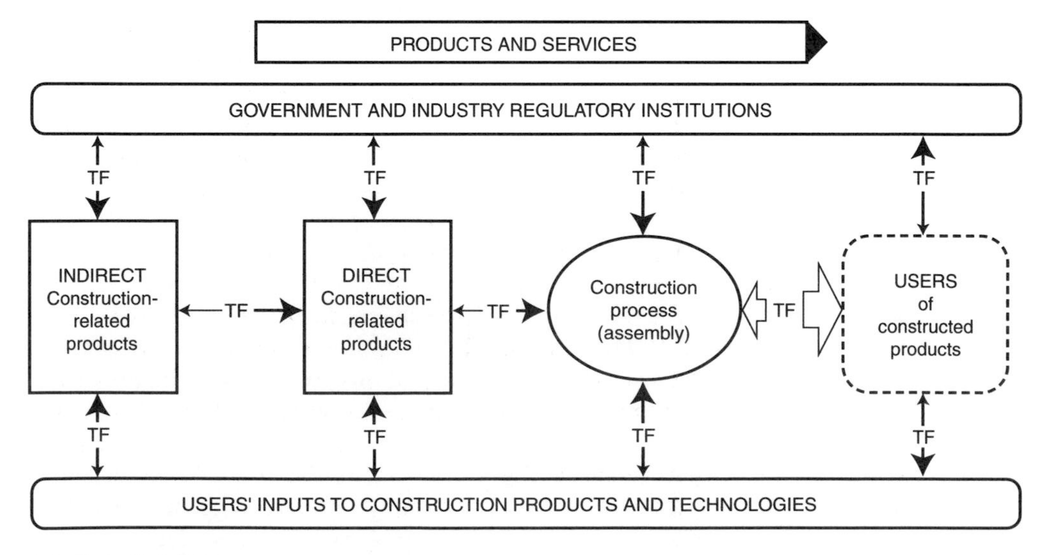

Fig. 7.3. Technology flow (TF) and commodity flow in the construction system (source: developed from Gann et al. (1992))

flows of technology, an approach is required that takes account of chains, streams and flows of production (*filières*), complexes of industries as well as their size, type and structure, and relationships within and between firms (Marceau 1992, pp. 467–469).

Innovation in construction can therefore only be properly understood through an examination of changes both on and off sites and at sectoral and industrial boundaries. The analysis needs to focus on the *construction system* as a whole, including inputs and outputs and the relationships between the many different participants — from clients and designers, manufacturers and suppliers of materials, components and equipment to contractors and specialist engineering firms. This way of viewing technological change has implications for Perez's concept of socio-institutional mismatches which must be adapted to explain mismatches that may occur in several stages and at different levels through the system. Such mismatches may result in new barriers to diffusion, as old ones are removed and the production system mutates and adapts.

(2) The relationships between producers and users require more detailed attention if problems arising from a poor understanding of user needs in the development and use of new technologies are to be avoided. The relationship between producers and users of the built environment is of particular importance given the nature of the construction system, with its myriad of supply networks. Much of the formal R&D carried out in the construction supply industries is oriented towards developing better products for building end-users, rather than for improving techniques in the construction process itself (Gann *et al.* 1992, p. 7). For example, in 1989, the UK construction supply industries spent about £267 million on R&D, of which around £214 million was indirectly 'embodied' in products which flowed through the construction process into buildings and structures, to benefit end-users. This reinforces the need to understand processes of technology transfer and the part played by construction as a conduit for technologies originating elsewhere and intended to benefit end-users. The danger is that poor information flows, lack of feedback and the distance between end-users and upstream suppliers may result in the development of inappropriate technologies.

Construction markets are segmented and they may be highly specialized. Large users with specialist requirements (e.g. those in the power and water industries, airport and aviation authorities, retail chains, or certain types of manufacturing firms) play a part in developing construction technologies to meet their own needs. These users often have their own teams of construction experts who work closely with others in the supply system to develop particular solutions. These types of relationships are similar to those found in some other industries, for example the production of scientific and medical instruments (von Hippel 1988, Ch. 2). User–producer relationships, which evolved through the need to meet specialized requirements, involve the transfer of know-how up and down the construction system: such projects are often at the forefront of technological change. Producers who work closely with users are likely to gain through the interactive learning processes that occur in the development of new technologies (Lundvall 1992, p. 48, Rothwell 1994). Nevertheless, many clients lack the expertise needed to interact with construction and its supply industries; furthermore, most are infrequent or once-only purchasers. They therefore often have to rely upon professional advice from an array of consultant design, engineering and construction organizations.

(3) The regulatory environment within which innovation takes place is often not considered in sufficient detail in the innovation literature published to date. Government plays a role which may both stimulate and retard the rate of technical change in construction. For example it is responsible for issues relating to the 'public interest', by setting legislation aimed at ensuring that buildings are constructed to a satisfactory quality to prevent people being killed or injured by poor design and workmanship, the spread of disease or fire, and degradation of the environment. In this role, government agencies are acting as proxies for users. Statutory duties are laid down in building regulations and codes, and practices are developed to encourage good design and construction techniques (discussed in Section 8.2). This activity usually involves incremental changes through revisions to existing legislation and codes. It can also slow down the introduction of radically new technologies that do not comply and which would entail major new regulations and standards before they could be used. National and local governments also have some control over the types and locations of buildings through their involvement in planning.

(4) Not enough is known about the processes by which know-how is acquired and passed on to others. This is important in explaining the rate and direction of technological change, particularly in construction where

knowledge is often transferred informally. Traditional approaches to the transfer of know-how assume continuity in learning, technology and theory and practice. This level of continuity ensures that, for the most part, construction technologies evolve slowly through minor changes and adaptations. But continuity is thrown into question when radical technical change occurs. New core competencies are required to exploit these opportunities and the knowledge base changes through a variety of different learning processes, some of which are discussed in Section 8.1.

(5) Firms are beginning to position themselves in new markets to deliver integrated systems solutions in which services are becoming an increasingly important part of their activities. A better understanding of innovation in design, engineering, construction and operational services is needed, particularly in areas where value appears to be increasingly added through the provision of enhanced services associated with the use of complex product systems.

7.4. Summary

When radically new technologies evolve, it is likely that a series of mismatches occur between technological possibilities and current social and institutional capabilities to exploit them. A number of different social and institutional adjustments are usually necessary before new technologies can be successfully used. Construction invests comparatively little on R&D and an increase in expenditure would help the sector develop core capabilities in developing and managing technology. But other mechanisms of acquiring new technologies are also important in this sector, such as technology transfer and a variety of learning processes. Strategies are needed to strengthen these to enable firms and users to exploit the benefits from innovation.

The work of Freeman and Perez provides a schematic framework within which to analyse the effects of broad sweeping changes in techno-economic paradigm on the types of buildings produced and the construction techniques adopted. They highlight the importance of understanding the development of technology within particular social and institutional structures. In developing these ideas further from the general macro-economic level to a sectoral level, the work of Dosi proves to be helpful in explaining the evolutionary, irreversible nature of change. Moreover, the notion that the environment within which innovation takes place is *self-organizing* helps to clarify the evolutionary nature of the construction system. The order of the system emerges through dynamic interactions between technological progress, economic activities and strategies of firms and institutions governing decisions and expectations (Dosi and Orsenigo 1988, p. 21). The construction system is not an intentional, planned and controllable environment. Rather, it is one in which the introduction of new technology results in many competing organizational forms.

Two conclusions of significance to a general understanding of innovation in complex product systems can be drawn from this.

(1) Interactions in supply networks can be understood more clearly by focusing on technology flows through *systems of production*, rather than within discrete sectors. This provides a useful method for identifying bottlenecks due to mismatches in the adoption of technical innovations and existing socio-institutional structures, particularly where complex products are created from the inputs of industries from many sectors, and when technological change often occurs at sectoral boundaries.

(2) The 'supply network' approach is in itself insufficient unless it is related to the role played by users of complex systems whose needs emerge over time. Changes in complex systems are sometimes triggered by users' demands, such as requirements by financial services firms for new buildings to house their electronic dealing rooms. Further development and subsequent refinements are strongly influenced by users' needs, as was shown to be the case in the transition from first to second generation intelligent buildings. Moreover, these systems are adapted and added to over time as users' experiences grow and as their needs change. The systems do not remain static from the moment they are created—they undergo successive changes throughout their lifecycles. The development of better vintages therefore depends to a large extent on knowledge of the experience of systems in use.

It will only be possible to systematically improve the quality, cost and user-friendliness of the built environment through a thorough grasp of how the construction system works, its history, and the strengths and weaknesses of different organizational forms. Analysis of technological change needs to be situated within an analytical framework that defines construction as a system of production. A synthetic approach has been developed here, which attempts to build bridges between macro-level shifts in the nature of demand and micro-level analysis of firms' attempts to develop new systems. This could help policy analysts understand processes of industrial restructuring, identifying pressure points of change and developing innovation strategies at firm and industry levels. In particular, it can help in an understanding of the relationship between increasing product complexity and the need for new specialist and interdisciplinary skills. It may also assist in understanding how to improve the total process, in contrast to focusing too narrowly on trying to optimize sub-processes. By focusing on the changing system of production rather than a narrowly defined sector, this approach escapes the problems of viewing construction as a *static industry.* Once this is avoided, and construction is viewed as a *dynamic process*, it is possible to develop new policies for the management of change in these project-based activities—this is the subject of Chapter 8.

8. Managing innovation in project-based firms

Previous chapters have shown that during the past two decades the built environment has changed radically in OECD countries, with the need to accommodate new social and economic activities in the digital age. Leading design, engineering and construction firms have responded to the challenges of developing new types of buildings and structures. They have searched for ways to improve performance, reduce cost and waste and speed up production, whilst developing new approaches to whole-life costs and flexibility in the use of buildings. Some firms have attempted to improve quality by reducing defects. These firms also strive to increase their profitability. Yet evidence suggests that there is much room for further improvement and that pressures for product and process innovation will continue in the coming decades. Clients are demanding and expecting more from their suppliers and innovative firms are seeking new ways of delivering products and services.

Encouraging signs of improvement in knowledge about innovation in construction were evident in the 1990s in Europe, the USA, Canada and Japan. Structured approaches to promoting innovation emerged, involving collaboration between industry, government and research organizations. Systematic management of innovation was helping to close the gap in performance between construction and other industries. Nevertheless, contradictions continued to exist in the policy arena, flowing from the many conflicting interests represented in, and affected by, the production and renewal of the built environment.

Existing barriers to the ways in which we think about innovation need to be softened in order to unlock new approaches to managing change. Practices from the machine age have ossified within increasingly irrelevant discipline-based value systems that hinder the development and transfer of new knowledge. It is necessary to erode boundaries between professions that span a wide range of specialist interests from planning and development to design, engineering and management. A culture of innovation is needed in which people from different professional backgrounds work together in new ways, motivated by the aim of meeting emerging needs of users and improving performance of the construction sector as a whole.

This Chapter focuses on key management and policy issues for innovation in the built environment. It explores how firms can develop innovation strategies and new technical and organizational capabilities to compete successfully in the future, assessing what types of specialist technical skills and general management skills will be needed. It examines the roles governments can play in promoting the conditions for efficient and effective construction in which firms can compete successfully, whilst at the same time setting regulations to protect public interests and stimulate further performance improvements. These issues are addressed below in the context of a number of core themes discussed throughout this book. These themes include the need to develop new integrative competencies and rational–adaptive approaches to the management of innovation in complex buildings and structures. Of equal importance is the need to understand changing economic and social contexts in the management of technology in the built environment. Linkages need to be improved between project processes and within business processes in firms. Firms also need to embrace the provision of services to provide support for the use of facilities, transferring relevant knowledge from initial design and construction to operation of buildings and structures. The Chapter concludes with a speculative glimpse into the future of innovation in the built environment in the coming biomolecular age.

8.1. Project-based, service-enhanced firms

Four factors have come to the fore in determining success for innovative design, engineering, construction and supply firms involved in producing the built environment in the digital age. These factors are described below and then each is considered in more detail.

(1) As products become more complex, many firms need to develop new capabilities to deliver integrated systems solutions and services to enhance the value they provide to clients. Long-term changes in emphasis from products to services and from the structures and fabric of buildings to plant, equipment and control networks mean that project-based firms need to embrace new roles.

(2) Firms need to implement new mechanisms for managing learning processes and knowledge within their businesses. The ways in which knowledge is produced and used are changing rapidly in the digital age, representing new opportunities as well as threats to the delivery of traditional professional and operative skills. In particular, the use of information and communication technologies (ICTs) can provide the tools to assist in transferring information between projects and central resources

within firms, enabling them to develop new knowledge about production processes.

(3) Firms need to develop a better understanding of the changing balance between general and specialist skills needed to operate in the digital age. They need to devote more resources to developing their technical and management competencies, which are likely to be of increasing importance in differentiating them in the marketplace.

(4) Firms need to improve their capabilities in managing innovation and technology if they are to build reputations for technical excellence that set them apart from more traditional players. Leading design, engineering and construction firms have shown that they can improve their use of existing technologies and that this can lay the foundations for further technological innovation.

8.1.1. Integrated systems solutions and services

The traditional boundary between manufacturing and services is fast becoming obsolete as new forms of manufacturing emerge to supply physical products packaged together with intangible services. For example, rapid productivity growth in the USA during the 1990s has been in part attributed to the ability of manufacturers to resolve tangible, physical problems through the provision of a wide range of services (Lester 1998). At the heart of this agenda is a programme of investment in 'intangible' assets such as ideas, information, skills and organizational capacities, including the capability to understand differentiated and often conflicting customer requirements. ICTs provide the new infrastructure for this transformation in business processes, allowing many services to be stored and transported over long distances. These trends affect the production and use of the built environment.

Prime contractors, project management firms and large integrated design and engineering organizations face a new competitive challenge: to develop the capabilities to provide fully integrated systems and service solutions, providing customers with single point purchase and support for buildings and structures (Davies, A. and Brady, T. 2000). These solutions include new forms of turnkey projects such as build, own, operate and transfer (BOOT) arrangements and their variants. They include many types of private finance and outsourcing contracts for public sector facilities. These changes are associated with clients demanding higher levels of service from their suppliers throughout product lifecycles—from financing, design and systems integration, through implementation, to technical support, after-sales maintenance, operation and decommissioning. Adding value through the provision of integrated systems and services solutions involves the 'soft' process of interacting more closely with customers to understand their requirements. It also involves the development of 'hard' ICT tools to capture and feed back customer requirements into front-end systems integration activities.

In many cases such changes have resulted in the creation of new service-based organizations, sometimes in joint venture with other firms. These project-based, service-enhanced organizations also require new business management competencies, including whole-life financial accounting and capabilities in supply chain management. They play a pivotal role as systems integrators at the centre of the new production system, coordinating design, integration and assembly and facilitating feedback between users and producers up and down the supply chain, as shown schematically in Fig. 8.1.

Two other elements of integrated systems and services delivery are proving important. First, the need to manage different political interests and the tensions they cause in project development. Environmental concerns are such that these can no longer be treated as 'externalities' in project planning, and new approaches need to be developed to include participants from a wide cross-section of interest groups in early stages of decision making. Second, systems integrators need the ability to manage technical complexity, including an understanding of technical interfaces between sub-systems and organizational interfaces between work tasks in design and construction. They need to manage increasingly complex design environments in which it may be beneficial to provide longer design lead-in times and shorter production schedules, providing more time for planning and less time for implementation and making mistakes on sites. These requirements mean that it has become increasingly difficult to rely solely upon technical, engineered solutions. Indeed, interdisciplinary approaches that combine knowledge from both engineering and social sciences are needed.

8.1.2. Learning and knowledge

In manufacturing sectors, the ability to manage technology has been shown to be a critical success factor of innovative firms. In

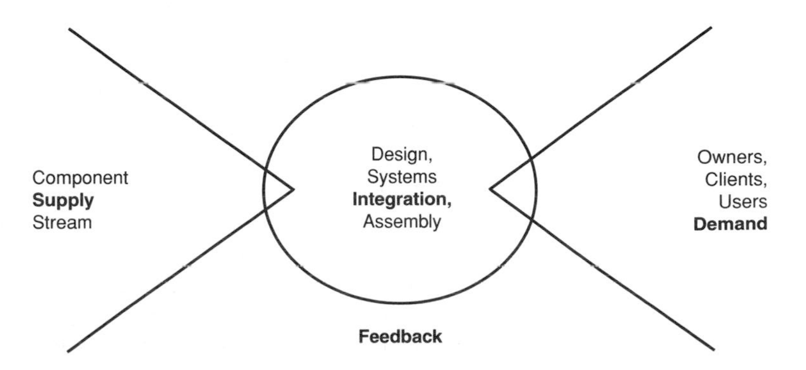

Fig. 8.1. The integrated production system

some areas, technical know-how is becoming more important than capital resources as a foundation for success (Leonard-Barton 1995). Yet few design and construction organizations have explicit policies for knowledge management, although they are having to adjust to new business environments in the digital age. At the same time, sources and means of access to knowledge itself are changing.

It is likely that the body of knowledge relating to building and construction science and management is expanding faster than the rate of increase in resources for research and training. Expansion is occurring in two dimensions—in the breadth of knowledge measured by the number of new specialist disciplines required to produce buildings and structures, and in the depth of knowledge measured by the extent of education and training needed to practice in particular specialisms. Firms' capabilities to absorb new knowledge at a time when knowledge sets are expanding along these dimensions are being stretched. For these reasons, companies need to develop strategies for mobilizing resources to move into rapidly emerging areas, such as electronic building controls, integrated systems and services solutions.

A radical shift in the production of knowledge has been occurring in industrial and consulting organizations, universities and government since the mid 1980s, affecting the types of knowledge developed to support innovation in the built environment. These changes have been described as a shift from Mode 1 to Mode 2 knowledge production (Gibbons *et al.* 1994).

Traditional knowledge production systems (Mode 1) were based on a clear demarcation between the public and private sectors. Universities were independent, discipline-based providers of basic education and skills to students. They carried out research which they believed was in the long-term public interest of adding to 'the body of knowledge' within a particular discipline. Much of this knowledge was produced with the intention that it should be used by other academics. Quality was scrutinized and controlled within well defined frameworks such as peer review of research proposals and refereed journal articles. Consultants and other organizations provided knowledge to private organizations and government.

Under this regime, technology transfer from universities to design and construction firms occurred through a number of mechanisms, including apprenticeships, peer group learning, the simulation of practice, participant observation, didactic teaching and learning and personal or directed study (Groák 1992, p. 163). These mechanisms provided the traditional means by which design and construction practitioners learnt their trades and disciplines during the machine age.

Within the emerging context of Mode 2, distinctions between public and private knowledge production have become blurred. Universities are involved in consultancy and industry is a significant participant in academic research, development and training. Knowledge production has shifted towards inter-disciplinarity, research in the context of application, problem solving and greater collaboration. At the same time, the quality of outputs has become more varied and is difficult to measure. The distinctions between Mode 1 and Mode 2 are shown in Table 8.1.

Some leading firms are embracing Mode 2 knowledge processes in order to enhance their organizational and management practices, with the specific aim of improving their capacity to innovate. These firms recognize that new in-house technical competencies are required in order to benefit from their own research and absorb results of research carried out elsewhere. As a consequence, they are developing new approaches to learning and feeding back knowledge from one project to the next.

The development of intra-firm Intranets provides a new mechanism to support the dissemination of good practices and feedback within firms. In theory, Intranets offer the opportunity for firms to improve their management of knowledge by allowing individuals to share internal information more easily. Yet firms in the construction sector often struggle to fill their Intranets with useful information. Usage patterns of Intranets vary considerably among these firms and, in order to increase usage rates, some have launched online journals, enabling project staff to share experiences. The rapid rate of development and diffusion of these technologies within leading firms in many

Table 8.1. Modes of knowledge production from the machine age to the digital age (source: adapted from Gann and Salter (1999), originally from Gibbons et al. (1994))

Mode 1, machine age	Mode 2, digital age
• Discipline-based teaching • Clear demarcation between universities and industry, academic and consultant • Universities educated, industry trained • More students means a better education system • Peer review • High levels of trust in science—independence	• Interdisciplinary learning • Blurring boundaries between universities and industry, academic and consultants • Greater collaboration • Knowledge production widespread in society • Learning organizations • Research in the context of application • Declining trust in science, scientists and engineers

industrial sectors suggests that they will provide powerful means by which knowledge flows can be improved during the next decade.

In this turbulent environment, design, engineering and construction organizations need to assess their core capabilities, understanding how these are deployed and how they can be renewed. They need to develop mechanisms for capturing lessons from projects, as well as from markets, clients and users, and from other sources outside the firm. Most importantly, they need to develop better mechanisms for sharing information and knowledge internally.

8.1.3. General and specialist skills

Innovation creates the need for new competencies, sometimes destroying older capabilities. But in the production and renewal of long-lived buildings and structures, there is often a need to maintain many older skills from the machine age, augmenting these with newer digital age capabilities. At the operative level, on sites, the general trend is away from manual, mixing and cutting skills based on traditional crafts or specialized machine age processes. The new requirement is for skills in measuring, alignment and fixing, together with a better understanding of the technical aspects of work associated with handling materials and components and managing work packages. Many small specialist contractors lack the skills to introduce new technologies, particularly in data-cabling, opto-electronics, controls, switchgear, and building management systems installed within intelligent buildings. There are also serious problems with operative skills in cross-functional areas (such as maintenance) in which newer technologies are used in conjunction with older systems.

Changes in the digital age are also creating new demands on professionals and managers working in the built environment. In many respects their jobs are becoming more exacting as buildings and structures become more complex. Two interrelated trends are evident in skill requirements at all levels of work in the production and maintenance of the built environment. Specialization is increasing, particularly where new technical skills are required. This in turn creates requirements for an increasing number of generalists who are capable of working across traditional disciplinary boundaries. The trend towards specialization creates tensions between the skills and activities of individuals and the need for integration and coordination among various actors. Coordination skills depend upon understanding the activities of others and an ability to integrate and manage different sets of knowledge.

Two types of interdisciplinary skills are needed—those of a technical nature, which assist in coordination between different technical specialisms and those which span technical and social science areas, such as economics, law, communications and politics and cultural analysis. These skills often complement the emerging patterns of specialization, providing the means to make connections across disciplines so that processes of design and construction can become seamless. They are particularly important in solving problems in areas where there are a large number of variables and high uncertainty and risk, such as in many complex building and construction projects. Systems integration capabilities are required, especially where new technologies need to be installed and operated within older technical systems.

New interdisciplinary and specialist skills are also needed from a business management perspective, particularly where business processes within firms need to be coordinated alongside the project processes in which firms are working (Gann and Salter 1999). Some of the main trends in new professional skills are shown in Table 8.2. These skills are not an either/or. Both are essential.

The traditional role of construction professionals as 'knowledgeable experts' requires augmenting with additional interdisciplinary skills to enable participation as part of a team. The model of 'expert advisers' providing discipline-based knowledge at certain times in the total production process is beginning to be challenged by new ways of organizing processes in which systems integrators and 'knowledgeable team players' are required. In this mode of working, construction professionals operate as 'practitioner–researchers', renewing their knowledge through participation in research and innovation activities, as well as through their daily work on projects (Groák and Krimgold 1989).

Design is becoming increasingly important in determining competitiveness because it embodies the knowledge to relate user requirements with technical possibilities. More value is being added in design and engineering processes in the production of complex products, often in collaboration with systems and component suppliers upstream in supply chains. Design skills are themselves becoming specialized within particular areas of capability, such as in assessing fitness for purpose, visual appearance, user-friendliness, efficiency in production and use of materials, safety, environmental impact and durability. Yet most design work involves tacit knowledge. Design remains an intuitive, creative process, the practice of which may be difficult to understand and test empirically. Design activities are therefore usually poorly measured both within firms and across the construction sector as a whole. However, there are increasing pressures to develop scientifically verifiable approaches to design and this is leading to firms developing new general capabilities in integrated design and design management.

Table 8.2. Trends in new general and specialist skills

New specialist skills	New general and integrative competencies
• Brief development and definition	• Environmental planning
• Design management	• Transport planning
• Production planning, assembly and installation management	• Space planning, syntax and changing working patterns
• Specialist project finance	• Business analysis
• Specialist legal advice	• Dynamics and complex systems analysis
• Risk assessment and management	• Building economics and lifecycle analysis
• Safety management	• Team building, co-locating, concurrent engineering
• Supply chain management	• Partnering and supply chain management
• Procurement and logistics	• Interdisciplinary skills to integrate engineering and social science expertise
• Instrumentation and control systems	• Understanding users and regulatory frameworks
• Non-destructive testing	• Delivery of integrated products and services
• Facilities management	
• Energy management	
• Water management	
• Building physics	
• Materials science	
• Contaminated land engineering	
• Geotechnical engineering	
• Structural engineering	
• Façade engineering and design	
• Mechanical and electrical engineering	
• Heating, ventilation and air conditioning	
• Manufacturing engineering	
• Wind, seismic and vibration engineering	
• Fire engineering	
• Lighting design	
• Acoustical engineering	
• Simulation and modelling	
• Computational fluid dynamics	
• IT systems and data management	
• Documentation control	
• Machinery operation and maintenance	

8.1.4. Managing innovation

Project-based construction firms often struggle to learn from project to project. Whilst they may have strong capabilities in project management or particular technical specialisms, they are often much weaker in organizing their internal business processes. This means that projects are often viewed as one-off and task-oriented. This often promotes a culture in which each project is approached from a new angle, limiting the extent to which feedback and learning takes place from one project to the next. Even when a firm has the technical competencies to absorb new ideas, it may not have the internal structure, systems or cultural attributes necessary to capitalize on them.

Firms need particular skills and an appropriate infrastructure in order to innovate and manage technical capabilities. Three key features characterizing successful innovators are: strong leadership, competent interdisciplinary people capable of working in teams, and an appropriate infrastructure to support implementation. A blueprint for managing innovation in construction organizations was set out in *Profit for Innovation*; the ideas in this document have since proven successful in practice within a number of leading UK design and construction organizations (see CIC 1993, Gann and Simmonds 1993). The possibility of developing innovative capabilities without incurring large costs has been demonstrated. For example, firms wishing to improve their performance through the use of new technologies have succeeded by employing innovation directors, technology managers, gatekeepers and facilitators to coordinate the use of existing and often latent technical know-how. The typical responsibilities of an innovation director are to:

- formulate the innovation strategy and update it at regular intervals, auditing the firms' technical capabilities and identifying key 'hot spots' (problem areas) and 'beauty spots' (areas of excellence) within the firm;
- act as the guardian and champion of the strategy and prioritize initiatives, including incremental and more radical improvements;
- prepare financial plans for development and implementation against clearly defined targets;
- establish a timetable for activities and milestones and metrics for assessment of progress;
- manage information flows between general management, technologists, engineers and designers;
- promote an environment conducive to innovation, nurturing and mentoring innovators, providing space for them to make mistakes and learn from these experiences;
- identify training needs associated with new practices and technologies within the firm;
- communicate with users and other interested parties within the firm about possibilities for innovation and encourage suggestions for improvements from people at all levels. Publish newsletters, Intranet bulletins and case study reports, encouraging feedback;
- develop a knowledge bank of key technical and operational factors, which others in the firm can draw upon in support of performance improvement;

- champion the implementation of new ideas with potential external users, clients and suppliers, obtaining feedback from them;
- exploit relevant external expertise, fostering relationships with research clubs, universities and other innovative firms;
- develop a foresight capability to enable assessment and exploitation of emerging technological and organizational opportunities and to help align the innovation strategy with possible future shifts in the market and the needs of users;
- interact with government agencies, professional institutions and industry organizations in relation to technical standards, regulations and the need for training;

Many small design and construction organizations cannot afford to take large risks, yet faced with changing market, technological and business conditions, they need to innovate. Successful firms have started by understanding their existing core capabilities, together with their strengths and weaknesses. Their innovation strategies have been set against realistic objectives, closely aligned to their capabilities and linked to known market opportunities. For example, it is usually helpful if a few, readily understandable key performance indicators are identified and used to set targets for performance improvement. These might include measuring repetitive defects in order to improve quality, or counting the number and cost of skips used per unit in order to stimulate ways of reducing waste on sites, or measuring delivery times to improve supply chain coordination. Firms like Willmott Dixon, Wilcon and Westbury in the UK, and Neenan Construction or Doyle Wilson in the USA, have been able to develop innovation strategies from these starting points. Risks have been minimized by prototyping new approaches off-line from real projects, or in isolated demonstration projects, often in collaboration and with support from clients.

The development of an infrastructure to support and sustain implementation is of crucial importance. This infrastructure could include a database, or knowledge bank of technical options, standard details, simulations and sources of expertise, etc. It could be made available on the firm's Intranet, thus enabling remote access to designers, engineers and managers working on projects, enhancing the possibilities of capturing and feeding back lessons from the field.

Larger construction organizations that already have formal technical support functions with responsibilities for trouble shooting, problem solving and R&D, have been able to improve their technical capabilities further through taking three steps to improve the management of innovation. First, they can map the ways in which technical resources are mobilized, fed back and developed within and between projects and central support services. Second, they can measure the types of activities carried out within formal R&D and technical support groups in order to identify better practices in the provision of future support, particularly using Intranet technologies. Third, they can develop better ways of utilizing and feeding back technical knowledge and information about processes of change with key supply firms and with long-term clients. Technological transfer usually occurs through interactions among people in business and those trained close to active researchers in universities, research institutions or other research departments. Innovative firms can therefore enhance their capacity to improve performance by co-locating staff in research organizations and with suppliers or clients.

Successful commercialization of research outcomes requires more than a straightforward linear mechanism for carrying out a precise programme of research. Firms need to be actively involved with researchers in universities and other research institutions to provide a research infrastructure, capable of developing and sustaining a knowledge base in new areas. The nature of these partnerships will prove particularly important in an industry where producers need to learn more about how to provide value to their users, and clients need to learn to become more demanding.

The operation of explicit, systematic innovation and technology management strategies enables firms to develop their reputations for technical excellence. This helps to differentiate them in the marketplace. In project-based environments, firms are often seen to be as good as their last project. They sell themselves on their reputations in specific areas. Loss of reputation correlates closely with loss of business. This leads to the need for vigilance in connecting activities at the project level with internal resources and business processes inside the firm. Firms' resources are bound up with their internal routines, they are the firm-specific assets of the organization. The use of resources for competitive advantage lies in their timeliness, value, rareness and costs of imitation (Teece and Pisano 1994). Developing and sustaining reputations for technical excellence and the capacity to innovate is therefore an important senior management function.

The resources of the firm are embedded at both the project and the firm level. It is the integration of these two sets of resources that enables the firm to be competitive. Business processes are ongoing and repetitive, whereas project processes have a tendency to be temporary and unique. Firms usually routinize their business activities and such routinization is made possible by their very nature (recurrence and frequency). Routines can stimulate innovation, providing opportunities for sustained process improvements. By contrast, project processes usually present unique features that do not lend themselves to systematic repetition. This can limit opportunities for process improvement, standardization and economies of scale.

A model indicating the interactions within which technical support and service activities are linked within business processes inside the project-based firm to project processes in the field is shown in Fig. 8.2. This illustrates in-house support and external research and technical support services bought in by firms. In general, knowledge associated with 'know-what' and 'know-why' tends to be codified, whilst knowledge associated with 'know-who' and to some extent 'know-how' tends to be uncodified, or tacit. Tacit knowledge is extremely important in this environment (particularly in the provision of management services) and it is therefore important that firms develop a convivial culture within which people are prepared to innovate in order to exploit new opportunities. To be successful, project-based firms need to integrate the experiences of projects into their continuous business processes in order to ensure the coherence of the organization.

Firms that are able to develop innovation and technology management capabilities generally find that they can recognize and exploit benefits of technical change. They are therefore in a better position to invest in R&D and the new skills required to guarantee future success.

8.2. Government policy

Governments have a central role to play in the promotion and support of innovation in the production of the built environment. These responsibilities and duties take a number of forms in support of the public interest, such as the protection of workers, consumers and the environment, and in furtherance of industrial competitiveness and overseas trade. Fulfilling these roles requires answers to complex questions about the development and use of new technologies in the built environment and

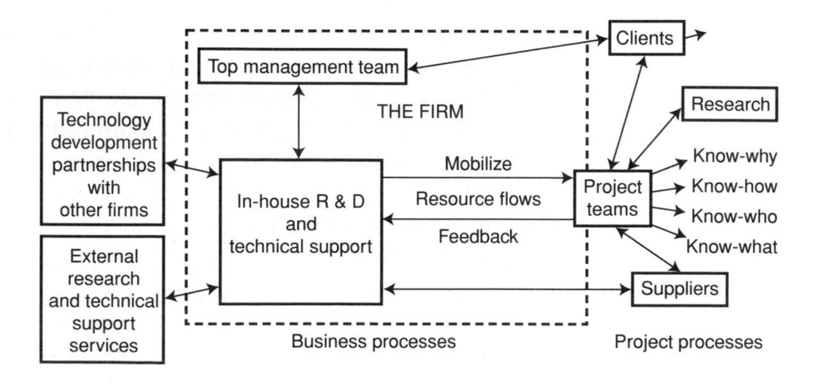

Fig. 8.2. Technical support resource flows

provision of information in the public domain about markets and technologies for which the private sector is unlikely to pay. It is also desirable for governments to focus attention on quality of life issues, for example the need for better pavements or the development of alternative solutions to existing housing or to road transportation. A range of direct and indirect policy instruments is available to governments, from fiscal and procurement policies to those in support of competitiveness and trade, R&D and regulation. A review of policy instruments for construction in Britain in the late 1990s is provided in work written up by Winch (1999).

The public sector's direct involvement in industry has diminished in many countries during the past two decades. As a consequence, governments have lost some of their capabilities to think about the future and plan support for technical innovation. In many countries, government policy makers work at some distance from hands-on decision makers in industry and they increasingly depend upon outside expertise concerning issues relating to technical change: knowledge about the production and adaptation of buildings and structures resides mainly within construction organizations, firms in supply industries, research and industry institutions. Moreover, governments often find themselves supporting contradictory policies—some aimed at developing competitiveness and others aimed at regulating practices.

Nevertheless, governments use a number of policy instruments that may affect the rate and direction of innovation. A range of trade and cross-border technology policies are used to provide support for the development and use of better construction technologies. Regulatory powers are increasingly being used to discharge duties, rather than direct government funding or ownership of facilities. For example, until around 1970, 50 per cent of all construction work was purchased by the public sector in the UK; this had fallen to less than 40 per cent by the mid 1990s. Governments nevertheless remain major customers for construction goods and services, for which they need to maintain knowledge about changes in demand and supply so that the public sector can be an 'expert' purchaser of goods and services. In Britain, shifts in policy towards private finance of public projects and new procurement policies aimed at improving overall value for money rather than lowest initial cost represent radical changes in procurement practices. Procurement policies are thus used as an important instrument to stimulate performance improvements through technical and organizational change.

In some countries, such as Japan, the UK and the USA, government departments have been involved in promoting a capability to understand future markets and technologies through technology foresight activities. These are important in order to guide choices about the investment of scarce development resources. Market insights are particularly important for firms and governments in demand-induced sectors such as the built environment, because the choice of technology is likely to relate closely to future user needs.

Government policies are important in areas where there is market failure in the development of technical know-how and the competencies needed to improve performance, such as within small and medium sized enterprises. Skills and training policies are a case in point. The transition from the machine age to the digital age has been accompanied by the need to develop a new portfolio of design and construction skills. OECD countries could benefit from better construction skills and, in every case, training programmes need to be modernized and working practices changed in order to develop a capability for working with new technologies (Gann and Senker 1998). In most cases the fragmented nature of construction with a wide range of representative bodies means that government needs to play a part in overall coordination.

Support for science and technology to develop public knowledge, competencies and technological capabilities has always been important in areas where there is perceived to be market failure on the part of the private sector. For example, few firms are likely to be able to appropriate the results of upstream science such as that carried out in new materials or computational sciences and other developmental work required to underpin knowledge for innovation in the built environment. Moreover, if this were to occur, appropriation by a small number of very large firms would almost certainly reinforce weaknesses in technical competence of the vast majority of small and medium sized enterprises. There is therefore a need for a strong, coherent and cooperative research base for construction. It is unlikely that this can be fostered solely within the private sector in most countries—with the possible exception of Japan—because of the structure and competitive nature of industry (Gann 1997). Governments therefore continue to support the development of design and construction technologies and new processes by funding R&D and promoting best practices across industry. In the digital age, with the technologies of knowledge production changing rapidly, policies for R&D need to address international, national and regional goals in relation to the type and size of firms and the nature of products and processes.

Applied research can be sponsored in collaboration with industry and research institutions through programmes that provide matching funding. Governments in different countries have pursued these roles through a number of policy instruments. Some (e.g. in Sweden) have used direct taxation and funding mechanisms. But the Swedish levy–grant system for funding construction research has tended to create a dependency culture in which industry expects the state to conduct research on its behalf. It has created a system in which research has become somewhat divorced from its users. In other countries, governments have facilitated construction research by acting as a catalyst to the development of new ideas; some of MITI's Programmes and those of the MoC in Japan have been successful in this respect. In other cases, governments have reduced taxes on corporate profits for those willing to invest in R&D. During the 1990s, the UK government launched a number of initiatives to stimulate collaborative research involving closer activities between industrial practitioners and university researchers. This resulted in the Engineering and Physical Sciences Research Council's Innovative Manufacturing Initiative and the Department of the Environment, Transport and Regions' Partners in Innovation programme. By contrast, the construction industry in the USA tends to be opposed to government intervention and has organized its own research clubs, such as the Construction Industry Institute in Texas, although for many years the military sector has provided a strong base of construction research through the US Corps of Engineers Research Laboratory. American firms have also made direct collaborative links with academics, for example in the Centre for Integrated Facilities Engineering at Stanford.

Perhaps of most concern in the long run is whether a competent cadre of research scientists and professionals capable of working with new technologies can be nurtured and sustained. Does government have a role to play in this? The answer is yes. There will always be a need to maintain a research capability as an insurance policy for meeting needs thrown up by uncertain events like the Ronan Point disaster in the 1960s and the impact of Chernobyl in the 1980s or the Kobe earthquake in the 1990s. Future technical problems will need to be resolved, associated with the use of new materials perhaps (similar to those experienced with the alkalization of concrete in the past) or the effects of acid rain on stonework. Other future research needs include knowledge about how facilities are used and the effects of buildings on users (e.g. sick building syndrome in intelligent buildings). In future, the biomolecular age will throw up a new set of requirements. Research capabilities are particularly important because constructed products last for a long time and it is not possible to predict changing patterns of use accurately. Moreover, the long life of products is set against peculiarly short time horizons of many of the project-based firms involved in their production, adaptation and maintenance.

8.2.1. Regulations

Three broad areas of regulatory policy are likely to affect innovation in the built environment: technical regulations affecting products and processes; planning and environmental regulations, principally affecting constructed products; and labour market regulations governing construction processes. There are also other legal provisions relating to different aspects of construction work, including regulations covering the provision of utility services such as gas, electricity, water and drainage.

The ways in which planning, building work and labour markets are controlled varies between countries. This is influenced by particular historical approaches within national legal systems, including laws relating to liability for defects in building structures, and different mechanisms for administering and supervising controls (see Institute of Building Control 1997). Building regulations were developed to protect the health and safety of occupants and those in close proximity to buildings. In most countries they rely upon standards and codes of practice covering structural safety, fire safety, health and hygiene, and safety in use. In addition, regulations often cover protection against undue noise, economy in the use of energy and, more recently, the provision of access and facilities for disabled people.

Constructed products are often highly visible to the public and their production and use can raise tensions between different interest groups, particularly when processes are dirty, dangerous or deplete the natural environment. The development and use of large complex buildings and structures is increasingly raising trans-national as well as regional and local issues about the governance of technology, for example in the case of air traffic control centres or cross-border infrastructural projects. Moreover, there is a need to comply with international agreements on the reduction of pollutants and emissions, such as those caused by heating and cooling buildings. These issues in themselves are creating the need for new regulatory structures and standards for building design and operation.

Regulations are viewed by many designers and builders as an additional burden to which they have to conform. For manufacturers, they set performance limits for components and materials. For clients, local authorities and government regulators, they provide the instruments with which product standards are maintained (see Gann *et al.* 1998). During the 1980s, there was widespread concern across many industrial sectors in Britain and the USA that too much regulation hindered competitiveness and new product development. This resulted in a backlash against all types of regulation, partly fuelled by an ideological belief in the ability of firms to deliver improved performance if they were left to operate in a free market (Porter 1990, pp. 647–649). The traditional prescriptive approach to building regulations came under scrutiny. This approach was criticized for being too rigid in application. For example, it often involved specification of the means by which regulations should be achieved, such as the materials, configurations and processes required, as well as the desired regulatory goal.

Whilst regulatory systems for the building sector vary between countries, by the 1990s there was increasing international convergence in approach. In particular, there was a strong preference for 'performance' rather than 'prescriptive' approaches. Performance regulations left the means by which regulatory goals were to be achieved open to practitioners, specifying only the final outcome. Performance regulations could also provide a basis for international comparison whilst accommodating national differences in constructed products and assessment methods. This was particularly important in the context of European harmonization during a time when regulations were increasingly becoming the focus of international competition and conflict (Sykes 1995, de Jonquieres 1998). By the mid 1990s, proponents of new performance-based approaches in Europe hoped that they would stimulate technological innovation. Their counterparts in the USA and Japan were following a similar course, based on the European approach.

It was recognized that regulations needed to accommodate technical change at different levels in the production process, including new product development and systems integration. Furthermore, regulations could be used to induce demand for high-performance emerging technologies, as was the case with energy regulations in buildings during the 1970s and clean technologies in the 1990s. More recent enthusiasm for performance-based building regulations focused on methods of conformity that consider buildings as whole systems, allowing trade-offs between component parts to achieve given objectives. This approach tended to stimulate systemic, holistic product innovation rather than purely innovation in component parts and sub-systems. Moreover, compliance mechanisms appeared to provide firms with the freedom, market incentives and institutional frameworks within which to innovate. The process of compliance itself could lead to greater information sharing and cooperation between firms. Nevertheless, firms needed to work hard under this regime in order to prove that innovative methods or new technologies met regulatory targets. The burden of proof sometimes slowed the rate of diffusion of innovations. This was particularly the case with techniques that appeared to work in practice but were difficult to prove with conclusive empirical evidence, for example in some geotechnical innovations. In other cases, monitoring complex buildings and structures proved costly and time consuming, and pre-production simulation models were therefore increasingly being substituted for post-production performance tests.

The onus of performance-based specifications is placed in the hands of the firms who have to implement them. This can increase operating costs and may be too expensive for individual firms to achieve compliance. There is not yet enough information available to assess these consequences, nor is it likely that such information will be collected within the private sector, and hence the need for research funded by government. However, it does appear that the regulatory process itself can stimulate change, leading to information sharing and cooperation between the public and private sectors, particularly if a portfolio of innovation and regulatory policies are developed. Competitiveness and regulatory policies therefore need to be carefully coordinated for this to be achieved.

8.3. Professional institutions and industrial organizations

In some countries like Britain, specialisms that emerged in the early stages of the machine age in the nineteenth century became institutionalized. Professional institutions and trade, or industry, associations emerged to perform a number of roles including protecting practices, institutionalizing quality control and acting as repositories of knowledge about success and failure in different types of projects. This latter activity proved particularly useful in a sector where discontinuities in feedback and information flows between projects and firms often resulted in failure to learn lessons about successes and failures. Professional institutions and industry organizations therefore often played an important part in maintaining and disseminating current knowledge about practices. Professional institutions also often helped to maintain standards by accrediting education programmes and providing their own professional training courses. As fields became more specialized they developed their own technical language and understanding. Research organizations were formed, often with assistance from government, in order to carry out testing of new components, develop standards and disseminate good practice.

Professional institutions publish journals for distribution to members. Debates about methods of practice, new technologies, regulations and government policies often take place in these publications. Technologists rely on these and the reports produced by research organizations to keep abreast of developments in the sector and their field of technology.

However, as technical specialization in construction increased, particularly with the rise of the digital age, the role of professional institutions and industrial organizations came into question. Many professional institutions guarded their domains, controlling the boundaries between different bodies of knowledge. Older professions, such as architecture, civil engineering and surveying, sometimes regarded newer professions like mechanical and electrical engineering or project management as a threat. This mirrored the division of labour in industry and often reinforced the breakdown in understanding between different players in design and construction processes. Many professions defended their traditional interests against newly emerging specialist and interdisciplinary practices. This hindered communications and the transfer of knowledge across disciplinary boundaries. By the 1990s, professions and industry associations were just as likely to cause inertia to change as they were to promote innovation.

8.4. The biomolecular age

The story thus far has shown how innovation in the built environment is linked to major changes in the economy and society, together with the introduction of new technologies during different periods in history. Most radical changes originated from outside the construction sector. Very often, new technologies and materials were first used in constructing buildings and structures to satisfy the requirements of the industries that had first developed them. During these periods of great change, product innovations, such as requirements to construct new types of buildings, generally took precedence over process innovations.

It is always difficult to envisage what the future holds, particularly in the middle of a long wave of diffusion and growth based on digital age technologies. Nevertheless, many futurologists and those involved in technology foresight exercises believe that we can already see the dawn of a new era. The next major changes, they argue, may come from biological engineering and associated new materials based on designing organisms at the molecular level (Ball 1997, Kaku 1998). This is the biomolecular age. The test will be whether these technologies can provide the foundations to underpin universal applications across other branches of industry by creating the possibility for new products, reducing costs in processes and enabling new services to be delivered. The signs are that biotechnologies and materials made to measure from molecular matter will achieve this.

Biotechnology could lead to new pressures for innovation in the built environment, again creating the need for new types of buildings, constructed from the materials of the biomolecular age. This could have as many consequences for the fabric and production of the built environment as those ushered in by the digital age. Microbiological organisms are already being developed for use in cleaning drainage systems and other equipment in buildings, such as in the removal of grease from drains associated with food preparation. Other biochemical materials are being developed for use in construction. Widespread adoption of these materials could result in the emergence of yet more technical specialists and at the same time pose difficult questions concerning systems integration, safety and durability.

In some technological visions of a world in which informational and biological sciences collide, our homes will need to provide spaces that can be programmed to support all types of activities and bodily functions. As Bill Mitchell puts it (1995, p. 105), new forms of interactive space might be needed to enable:

> *'bits to meet the body—where digital information is translated into visual, auditory, tactile, or otherwise perceptible forms, and conversely, where bodily actions are sensed and converted into digital information'.*

Biomedical technologies designed to serve an ageing population may result in some of these impacts.

In conclusion, the digital age will continue to challenge designers and constructors. The shift from centralized to decentralized networks both internally within many types of buildings and across organizations may increase. The need for flexibility in design and rational–adaptive approaches to management of construction will probably also grow. Distinctions between passive and active elements of buildings could become more blurred. New approaches in design and engineering, together with technological innovation in materials and structures, may tend towards the production of malleable and unstable structures, breathable and perhaps living skins.

Many people will continue to live in densely packed conurbations. For this reason, tall buildings are likely to remain important in some parts of the world and the work of designers like Yeang in developing new, bioclimatically considered approaches to skyscrapers may be of increasing importance (Yeang 1994). Such approaches may have spin-offs into other building types. Ecological and environmental issues are also likely to grow in importance and designers, engineers and constructors will need to develop new approaches to understanding and implementing 'sustainability' in buildings and construction processes.

Buildings and structures are generally long-lived and in most countries a large proportion of the built environment will need to be repaired, maintained and adapted well into the twenty first century. New types of materials and technologies may not displace many of the older ones in the built environment. Instead, the biomolecular age may serve to enlarge the palate of technologies used to produce buildings and structures. This immediately raises the issue of having to maintain skills from the machine and digital ages, whilst developing new capabilities for the biomolecular age. Future generations of designers, engineers and integrators will need to understand technical precedents from different eras, dealing with different layers of technologies.

This is a big challenge for a sector that often finds it difficult to plan resource requirements even a few months ahead. All those responsible for employing and educating people in the built environment will need to modernize their skills continuously. A culture of life-long learning is needed in industry. Failure to achieve this will leave a few more ghosts haunting the system, leading to a decline in quality, inflexibility, inefficiency and inappropriate use of resources. This will endanger our capability to design, construct and maintain the buildings and structures we need for a healthy, civilized and prosperous society in the twenty-first century.

Future prospects also present an exciting challenge, providing creative and stimulating opportunities for people working in new, interdisciplinary design, engineering, construction and management careers. It is of great importance that they delight in the new and provide continuity with the past, providing inspirational buildings and structures, enabling us all to lead fruitful and fulfilling lives.

References

Abercrombie, N., Hill, S. and Turner, B.S. (1986). *Sovereign Individuals of Capitalism*. Allen and Unwin, London.

Alarcon, L. (Ed.) (1997). *Lean Construction*. A.A. Balkema, Rotterdam.

Architects' Journal (1981). Brick supplement. Special Publication.

Arnold, E. and Gann, D.M. (1995). *Evaluation of IT-Bygg: The Swedish National Programme on Construction IT*. Technopolis, Brighton.

Atkins, W.S. (1994) *Strategies for the European Construction Sector*. Office for Official Publications of the European Communities, Brussels.

Atkinson, G. (1993). A view of the regulator's perspective. In: *The Responsible Workplace* (Eds F. Duffy, A. Laing and V. Crisp). Butterworth-Heinemann, Oxford.

Baldwin, J. (1996). *Bucky Works: Buckminster Fuller's Ideas for Today*. Wiley, New York.

Ball, M. (1988). *Rebuilding Construction*. Routledge, London.

Ball, P. (1997). *Made to Measure: New Materials for the 21st Century*. Princeton University Press, Princeton.

Banham, R. (1960). *Theory and Design in the First Machine Age*. MIT Press, Cambridge, MA.

Banham, R. (1969). *The Architecture of the Well-Tempered Environment*. Architectural Press, London.

Barlow, J., Cohen, M., Jashapara, A. and Simpson, Y. (1997). *Towards Positive Partnering. Revealing the Realities in the Construction Industry*. Policy Press, Bristol.

Barlow, J., Coupland, A., Marsh, G. and Morley, S. (1998). *Back to the Centre*. Royal Institution of Chartered Surveyors, London.

Barlow, J. and Gann, D.M. (1993). *Offices into Flats*. Joseph Rowntree Foundation, York.

Barlow, J. and Gann, D.M. (1995). 'Flexible Planning and Flexible Buildings? Re-using redundant office space'. *International Journal of Urban Affairs* **17**(3): 263–276.

Barras, R. (1995). The capital-saving city? *Proceedings of PICT International Conference on the Social and Economic Implications of ICTs*, Mimeo, Westminster, London.

Barras, R. (various dates). *Property Market Reports*. Property Market Analysis, London.

Bateson, K. (1999). Assault on 'battery farm'. *Building Design* **1389**: 17.

Bell, D. (1974). *The Coming of the Post-Industrial Society*. Heinemann, London.

Beniger, J.R. (1986). *The Control Revolution*. Harvard University Press, Cambridge, MA.

Berman, M. (1982). *All That is Solid Melts into Air*, Verso. London & New York.

Bernstein, H.M. and Lemer, A.C. (1996). *Solving the Innovation Puzzle: Challenges Facing the U.S. Design and Construction Industry*. ASCE Press, New York.

Billington, N.S. and Roberts, B.M. (1982). *Building Services Engineering—A Review of its Development*. Pergamon Press, Oxford.

Bordass, W. (1993). Building performance for the responsible workplace. In: *The Responsible Workplace* (Eds F. Duffy, A. Laing and V. Crisp). Butterworth-Heinemann, London.

Bottom, D., Gann, D., Groák, S. and Meikle, J. (1996). *Innovation in Japanese Prefabricated Housebuilding Industries*. CIRIA, Special Publication **139**, London.

Bowley, M. (1960). *Innovation in Buildings Materials*. Gerald Duckworth, London.

Bowley, M. (1966). *The British Building Industry*. Cambridge University Press, Cambridge.

Brand, S. (1994). *How Buildings Learn*. Viking Penguin, Harmondsworth.

Brynjolfsson, E.T.W. and Hitt, L. (1996). 'Paradox lost? Firm-level evidence on the returns to information systems spending'. *Management Science* **42**(4): 541–555.

Callahan, R. (1962). *Education and the Cult of Efficiency.*, University of Chicago Press, Chicago.

Castells, M. (1989). *The Informational City*. Basil Blackwell, Oxford.

Castells, M. (1996). *The Rise of the Network Society*, Blackwell Science, Oxford.

CFR (1996). *The Funding and Provision of Research and Development in the UK Construction Sector 1990—1994*. Department of the Environment, London.

CIC (1993). *Profit from Innovation: a Management Booklet for the Construction Industry*. Construction Industry Council/IPRA, Brighton.

CICA (1990). *Building on IT for the 1990s*. Construction Industry Computing Association and Peat Marwick McLintock, Cambridge.

CIRIA (1998). *Adding Value to Construction Projects Through Standardisation and Pre-assembly*. Construction Industry Research and Information Association **176**, London.

Clark, C. (1960). *The Condition of Economic Progress*. Macmillan, London.

Clarke, L. (1992). *Building Capitalism*. Routledge, London.

Construction Taskforce (1998). *Rethinking Construction*, DETR/HMSO, London.

Cooke, M. (1910). *Academic and Industrial Efficiency*, Carnegie Foundation for the Advancement of Teaching Bulletin **5**.

Courtney, R. (1997). Building Research Establishment—past, present and future. *Building Research and Information* **25**(5): 285–291.

Cowell, R. (1997). Simulating the acoustic environment. In: *Arups on Engineering* (Ed. D. Dunster). Ernst and Sohn: 124–131, Berlin.

Daniels, P.W. and Bobe, J.M. (1990). *Planning for Office Development in the City of London and Adjacent Boroughs*. Services Industries Research Centre Working paper, Portsmouth Polytechnic, Portsmouth.

David, P.A. (1985). Clio and economics of QWERTY. *American Economic Review* **75**(2): 332–337.

Davies, A. (1994). *Telecommunications and Politics: the Decentralised Alternative*. Frances Pinter, London.

Davies, A.C. and Brady, T. (2000). 'Building organizational capabilities in complex product systems'. *Research Policy—Special Issue on Complex Systems and Services*. forthcoming.

de Jonquieres, G. Rules for the regulators. *Financial Times*, 2 March 1998.

de Sola Pool, I. (1977). *The Social Impact of the Telephone*. MIT Press, Cambridge, MA.

Debenham, Tewson & Chinnocks (1992). *Central Office Research*. Debenham, Tewson & Chinnocks Holdings, London, Spring 1992.

DEGW, Ove Arup & Partners and Northcroft (1998). *Intelligent Buildings in Latin America*. Phase 1 Report. Intelligent Buildings Research Ltd., London.

DEGW/Technibank (1992). *The Intelligent Buildings in Europe*. European Intelligent Building Group, London.

Department of Commerce (1998). *The Emerging Digital Economy*. The U.S. Government, Washington, DC. April.

Department of Trade and Industry (1998). *Our Competitive Future: Building the Knowledge Driven Economy*. The Stationery Office Limited, London.

DETR (various dates). *Housing and Construction Statistics*. HMSO, London.

Devine, W.D. (1983). 'From shafts to wires: historical perspectives on electrification'. *The Journal of Economic History* **XLIII**(2): 347–372.

Dickinson, R.L. (1914). 'Taylorism in surgery'. *Journal of the American Medical Association* **LXIII**(9).

Dosi, G. (1982). 'Technological paradigms and technological trajectories'. *Research Policy* **11**: 147–162.

Dosi, G., Freeman, C., Nelson, R., Silverberg, G. and Soete, L. (1988). *Technical Change and Economic Theory*. Frances Pinter Publishers, London.

Dosi, G. and Orsenigo, L. (1988). Industrial structures and technical change. In: *Innovation, Technology and Finance* (Ed. A. Heertje). Basil Blackwell, Oxford.
Drewer, S. (1990). 'The international construction system'. *Habitat International* **14**(2/3): 29–35.
Duffy, F. (1997). *The New Office*. Conran Octopus Ltd., London.
Duffy, F. and Henney, A. (1989). *The Changing City.* Bulstrode Press, London.
Duffy, F., Laing, A. and Crisp, V. (Eds.) (1993). *The Responsible Workplace*. Butterworth-Heinemann, Oxford.
Dunsheath, P. (1962). *A History of Electrical Engineering*. Faber and Faber, London.
Edwards, K. (1998). *Technological Change in the Elevator Industry.* MSc Thesis, SPRU, University of Sussex, Brighton.
EPSRC (1994). *Innovative Manufacturing—A New Way of Working*. Engineering and Physical Sciences Research Council, Swindon.
ETSU (1987). *Energy Management Systems*. Energy Efficiency Office, London.
European Commission (1995). *Report of the Group of Independent Experts on Legislative and Administrative Simplification*. European Commission **288** Final/2.
Fazio, P., Moselhi, O., Thergerge, P. and Revay, S. (1988). Design impact of construction fast-track. *Construction Management and Economics* **5**: 195–208.
Feynman, R. (1960). There's plenty of room at the bottom. Reproduced in: May, A.S.G (Ed.) *Feynman and Computation* Perseus Books (1999) ch. 7: 63–76, Cambridge, Mass.
Finnimore, B. (1989). *Houses from the Factory: System Building and the Welfare State 1942–1974*. Rivers Oram Press, London.
Fitchen, J. (1986). *Building Construction Before Mechanisation*. MIT Press, Cambridge, MA.
Fitzpatrick, T. (1997). The balance of merit in designing tall buildings. In: *Arups on Engineering* (Ed. D. Dunster). Ernst and Sohn: 166–171, Berlin.
Flanagan, R., Norman, G. and Worral, H. (1995). Trade performance of the UK building materials and components industry. *Engineering, Construction and Architectural Management* **2**(2): 141–163.
Forster, B. (1997). Lightweight structures. In: *Arups on Engineering* (Ed. D. Dunster). Ernst & Sohn: 110–117, Berlin.
Forty, A. (1986). *Objects of Desire*. Pantheon Books, New York.
Freeman, C. (1978). Innovation and the construction industry. SPRU, MIMEO University of Sussex, Brighton.
Freeman, C. (1987). Information Technology and change in techno-economic paradigm. In: *Technical Change and Full Employment* (Eds C. Freeman and L. Soete). Basil Blackwell, Oxford.
Freeman, C. (1991). Innovation, changes of techno-economic

paradigm and biological analogies in economics. *Revue Economique* **42**(2): 211–231.

Freeman, C. (1994). 'The economics of technical change'. *Cambridge Journal of Economics* **18**: 463–514.

Freeman, C., Clark, J. and Soete, L. (1982). *Unemployment and Technical Innovation*. Frances Pinter, London.

Freeman, C. and Perez, C. (1988). Structural crises of adjustment: business cycles and investment behaviour. In: *Technical Change and Economic Theory* (Eds G. Dosi, C. Freeman, R. Nelson, G. Silverberg and L. Soete). Frances Pinter Publishers, London.

Freeman, C. and Soete, L. (1988). *Computer and Information Technology and the World Economy.* SPRU, University of Sussex, Brighton.

Freeman, C. and Soete, L. (1990). Fast structural change and slow productivity change: some paradoxes in the economics of information technology. *Structural Change and Economic Dynamics* **1**(2): 225–242.

Freeman, C. and Soete, L. (1997). *The Economics of Industrial Innovation*. Frances Pinter, London.

Gann, D.M. (1991). Buildings for the Japanese information economy. *Futures* June: 469–481.

Gann, D.M. (1992). *Intelligent Building Technologies: Japan and Singapore*. Department of Trade and Industry Visit Report, SPRU, University of Sussex, Brighton.

Gann, D.M. (1996). Construction as a manufacturing process? Similarities and differences between industrialized housing and car production in Japan. *Construction Management and Economics* **14**: 437–450.

Gann, D.M. (1997). Should governments fund construction research?. *Building Research and Information* **25**(5): 257–267.

Gann, D.M. (1999). *Flexibility and Choice in Housing*. The Policy Press, Bristol.

Gann, D.M. and Barlow, J. (1996). Flexibility in building use: the technical feasibility of converting redundant offices into flats. *Construction Management and Economics* **14**: 55–66.

Gann, D.M., Barlow, J. and Venables, T. (1999). *Digital Futures—A Critical Review of Specification, Implementation and Use of Smart Home Technologies*. Chartered Institute of Housing, Coventry.

Gann, D.M., Hansen, K., Bloomfield, D., Blundell, D., Crotty, R., Groák, S. and Jarrett, N. (1996a). *Information Technology Decision Support in the Construction Industry: Current Developments and Use in the United States*. Department of Trade and Industry, London.

Gann, D.M., Matthews, M., Patel, P. and Simmonds, P. (1992). *Construction R&D*. IPRA/Department of the Environment, London.

Gann, D.M., Phillimore, P., Thorburn, L., Simmonds, P. and Staton,

M. (1996). *Over-Specification of Engineering Services*. Technopolis, Brighton.

Gann, D.M. and Salter, A. (1998). Learning and innovation management in project-based, service-enhanced firms. *International Journal of Innovation Management* **2**(4): 431–454.

Gann, D.M. and Salter, A. (1999). *Interdisciplinary Skills for Built Environment Professionals*. The Ove Arup Foundation: 47, London.

Gann, D.M. and Senker, P. (1993). Construction robotics: technical change and work organization. *New Technology, Work and Employment* **8**(1).

Gann, D.M. and Senker, P. (1998). Construction skills training for the next millennium. *Construction Management and Economics* **16**: 569–580.

Gann, D.M. and Simmonds, P. (1993). *Profit from Innovation: Source Document and Research Report*. IPRA, Brighton.

Gann, D.M., Wang, Y. and Hawkins, R. (1998). Do regulations encourage innovation?—The case of energy efficiency in housing. *Building Research and Information* **26**(5): 280–296.

Gates, B. (1999). *Business @ the Speed of Thought*. Penguin Books, London.

Gibbons, M., Limoges, C., Nowotny, H., Schwartzmann, S., Scott, P. and Trow, M. (1994). *The New Production of Knowledge: The Dynamics of Science and Research in Contemporary Society*. Sage, London.

Gibson, G. (1998). Chapter 14, Nonconventional elevators, special applications, and environmental considerations. In: *The Vertical Transportation Handbook* (Ed. G.R. Strakosch). John Wiley and Sons: pp. 353–417, New York.

Giedion, S. (1967). *Space Time and Architecture*. Oxford University Press, London.

Gordon, J.E. (1976). *The New Science of Strong Materials*. Penguin Books, Harmondsworth.

Graham, S. and Marvin, S. (1996). *Telecommunications and the City: Electronic Spaces, Urban Places*. Routledge, London.

Groák, S. (1992). *The Idea of Building*. E. & F.N. Spon, London,.

Groák, S. (1998). 'Representation in building'. *RSA Journal* **CXLVI**(5487): 50–59.

Groák, S. and Krimgold, F. (1989). The practitioner-researcher in the building industry. Building Research and Practice, **17**(1) 1989: 52–59.

Guthrie, J. (1998). Builders spend 10% of turnover on tenders. *Financial Times*, 23 October 1998.

Habraken, J.N. (1972). *Supports—An Alternative to Mass Housing*. The Architectural Press, London.

Habraken, N.J. (1998). *The Structure of the Ordinary—Form and Control in the Built Environment*. The MIT Press, Cambridge, MA.

Hammer, M. and Champy, J. (1997). *Reengineering the Corporation: a Manifesto for Business Revolution*. Harper Business, New York.

Hardyment, C. (1988). *From Mangle to Microwave*. Polity Press, Oxford.

Harris, E.C. (1988). *Selection of Services—A Vital Factor.* E.C. Harris & Partners, London,.

Harrison, A., Loe, E. and Read, J. (1998). *Intelligent Buildings in South East Asia*. E. & F.N. Spon, London.

Harvey, D. (1989). *The Condition of Postmodernity.* Basil Blackwell, Oxford.

Haustein, H.D. (1980). *Lighting Industry: a Classical Case of Innovation*. International Institute for Applied Systems Analysis, Laxenburh, Austria.

Hayes, R.H. and Wheelwright, S.C. (1984). *Restoring our Competitive Edge: Competing Through Manufacturing*, John Wiley & Sons, New York.

Hillebrandt, P.M. (1984). *Analysis of the British Construction Industry.* Macmillan, London.

Hobday, M. (1998). Product complexity, innovation and industrial organisation. *Research Policy* **26**: 689–710.

Hobday, M. and Rush, H. (1999). Technology management in complex product systems (CoPS)—ten questions answered. *International Journal of Technology Management* **17**(6): 618–638.

Hoj, J., Kato, T. and Pilat, D. (1995). Deregulation and privatisation in the service sector. *OECD Economic Studies* 1995/11(25): 37–74.

Horgen, T.H., Joroff, M.L., Porter, W.L. and Schon, D.A. (1999). *Excellence by Design—Transforming Workplace and Work Practice*. John Wiley & Sons, New York.

Hughes, T. (1983). *Networks of Power: Electrification in Western Society, 1880–1930*. John Hopkins University Press, Baltimore.

Hughes, T.D. (1985). Edison and electric light. In: *The Social Shaping of Technology* (Eds D. Mackenzie and J. Wajcman). Oxford University Press, Oxford.

Hughes, T.P. (1998). *Rescuing Prometheus*. Pantheon Books, New York.

Ingels, M. (1952). *Willis H. Carrier: Father of Air Conditioning*. Country Life Press, Garden City, US.

Institute of Building Control (1997). *Review of European Building Regulations and Technical Provisions*. The Institute of Building Control, Epsom, 28.

Jencks, C. (1988). *Architecture Today.* Academy Editions, London.

Johnson, B. (1992). Institutional learning. In: *National Systems of Innovation* (Ed. B. A. Lundvall). Frances Pinter, London.

Josephson, M. (1961). *Edison—A Biography.* Eyre and Spottiswoode Publishers Ltd., London.

Kaku, M. (1998). *Visions—How Science will Revolutionise the Twenty-First Century.* Oxford University Press, Oxford.

KD/Consultants (1991). *Construction: a Challenge for the European Industry.* KD Consultants, The Netherlands.

Kendall, S. and Sewada, S. (1987). Changing patterns in Japanese housing. *Open House International* **12**(2): 7–19.

Key, T., Espinet, M. and Wright, C. (1990). Prospects for the property industry: an overview. In: *Land and Property Development in a Changing Context* (Eds P. Healey and R. Nabarro). Gower Press, Aldershot.

Kodama, F. (1992). Technology Fusion and the New R&D. *Harvard Business Review* July–August: 70–78.

Kose, S. (1997). Building Research Institute in Japan: past, present and future. *Building Research and Information* **25**(5): 268–271.

Kroner, W.M. (1989). 'Intelligent architecture through intelligent design'. *Futures* **21**(4): 319–333.

Latham, M. (1994). *Constructing the Team.* HMSO, London.

Leaman, A. and Bordass, B. (1997). *Productivity in Buildings.* The Workplace Comfort Forum, Central Hall, Westminster, London.

Leonard-Barton, D. (1995). *Wellsprings of Knowledge.* Harvard Business School Press, Boston.

Lester, R.K. (1998). *The Productive Edge.* W.W. Norton & Co., New York.

Linder, M. (1994). *Projecting Capitalism—A History of Internationalisation of the Construction Industry.* Greenwood Press, Westport, CT.

Lorch, R. (Ed.) (1997). What future for national building research organisations? *Building Research and Information, Special Issue* **25**(5).

Lorenzoni, G. and Baden-Fuller, C. (1995). Creating a strategic centre to manage a web of partners. *California Management Review* **37**(3): 146–163.

Lundvall, B.A. (Ed.) (1992). *National Systems of Innovation.* Frances Pinter, London.

Marceau, J. (Ed.) (1992). *Reworking the World.* Walter de Gruyter, Berlin.

Marceau, J., Manley, K. and Sticklen, D. (1997). *The High Road or the Low Road? Alternatives for Australia's Future.* Australian Business Foundation, Sydney.

Marsh, P. (1998). Management: AMP staff's efforts to keep in touch with customers and with each other. *Financial Times* 14 January 1998: 12.

Massey, D. and Meegan, R. (1982). *The Anatomy of Job Loss.* Metheun, London.

McCutcheon, R. (1975). Technical change and social need: the case of high-rise flats. *Research Policy* (4): 262–289.

Merrett, S. (1979). *State Housing in Britain.* Routledge, London.

Mitchell, W.J. (1995). *City of Bits: Space, Place and the Infobahn.*

The MIT Press, Cambridge, MA.
Mitchell, W.J. and McCullough, M. (1995). *Digital Design Media.* Van Nostrand Reinhold, New York.
Morgan, K. and Sayer, A. (1988). *Microcircuits of Capital: Sunrise Industry and Uneven Development.* Polity Press, Oxford.
Morris, P. (1994). *The Management of Projects.* Thomas Telford Services Ltd., London.
Morrison, C.J. (1997). Assessing the productivity of information technology equipment in US manufacturing industries. *Review of Economics and Statistics* **79**: 471–481.
Mumford, L. (1961). *The City in History.* Secker and Warburg, London.
Nabarro, R. (1990). The investment market in commercial and industrial development: some recent trends. In: *Land and Property Development in a Changing Context* (Eds P. Healey and R. Nabarro). Gower, Aldershot.
Nam, C.H. and Tatum, C.B. (1988). Major characteristics of constructed products and resulting limitations of construction technology. *Construction Management and Economics* **6**: 133–148.
NEDO (1978). *How Flexible is Construction?* Building and Civil Engineering Economic Development Council, National Economic Development Office, HMSO, London.
Negroponte, N. (1995). *Being Digital.* Vintage Books, New York.
Noble, N. (1997). The interactive building facade of the future. In: *Arups on Engineering* (Ed. D. Dunster). Ernst & Sohn: 96–103, Berlin.
Nowak, F. and Harling, K. (1996). *Evolution of Construction Process Research.* Building Research Establishment: p. 100, Watford.
Nye, D.E. (1994). *American Technological Sublime.* MIT Press, Cambridge, MA.
O'Brien, T. (1997). Materials Science in Construction. In: *Arups on Engineering* (Ed. D. Dunster). Ernst & Sohn: 66–75, Berlin.
OECD (1991). *Technology and Productivity—The Challenge for Economic Policy.* Organisation for Economic Cooperation and Development, Paris.
OECD (1997). National Accounts, 1983–1995. OECD, Paris.
OECD (1998). *Science, Technology and Industry Outlook.* OECD, Paris.
Pavitt, K. (1984). Sectoral patterns of technical change: towards a taxonomy and a theory. *Research Policy* **13**: 343–373.
Pavitt, K. (1999). *Technology Management and Systems of Innovation*, Edward Elgar, Aldershot.
Perez, C. (1983). Structural change and assimilation of new technologies in the economic and social systems. *Futures* (October): 357–375.
Perez, C. (1985). Microelectronics, long waves and world structural

change: new perspectives for developing countries. *World Development* **13**(3): 141–163.

Peters, T.F. (1996). *Building the Nineteenth Century.* The MIT Press, Cambridge, MA.

Petroski, H. (1996). *Invention by Design—How Engineers Get from Thought to Thing*. Harvard University Press, Cambridge, MA.

Pevsner, N. (1976). *A History of Building Types*. Thames and Hudson, London.

Porter, M.E. (1990). *The Competitive Advantage of Nations*. Macmillan, Hong Kong.

Proplan (1985). *Electronic Energy Management Systems in Buildings: the UK Market 1981–1990*. Proplan Search and Evaluation Services, London.

Rajan, A., Lank, E. and Chapple, K. (1999). *Good Practices in Knowledge Creation and Exchange*. Create, Tunbridge Wells.

Reckman, B. (1979). Carpentry: the craft and trade. In: *Case Studies on the Labour Process* (Ed. A. Zimbalist). Monthly Review Press, London.

Roach, S.S. (1991). Services under seige—the restructuring imperative. *Harvard Business Review* (September–October): 82–89.

Rolt, L.T.C. (1965). *Tools for the Job—A Short History of Machine Tools*. B.T. Batsford Ltd., London.

Rosenberg, N. (1976). *Perspectives on Technology.* Cambridge University Press, Cambridge.

Rosenberg, N. (1982). *Inside the Black Box*. Cambridge University Press, Cambridge.

Rosenberg, N. (1991). S&T Interfaces: critical issues in science and technology policy research. *Science and Public Policy* **18**(6): 335–346.

Rothwell, R. (1993). Issues in user–producer relations: role of government. *International Journal of Technology Management* Special Issue **9**(5/6/7): 629–649.

Russell, B. (1981). *Building Systems, Industrialisation and Architecture*. John Wiley & Sons, London.

Sauvant, K.P. (1986). *International Transactions in Services: The Politics of Transborder Data Flows*. West View Press, Boulder.

Schonberger, R. (1982). *Japanese Manufacturing Techniques*. Free Press, New York.

Schumpeter, J.A. (1976). *Capitalism, Socialism and Democracy*, George Allen & Unwin, London.

Scott Morton, S.M. (1991). *The Corporation of the 1990s*. Oxford University Press, Oxford.

Seaden, G. (1997). The future of national construction research organisations. *Building Research and Information* **25**(5): 250–256.

Senker, J. (1998). Turmoil in public sector building research—part of a wider problem. *Building Research and Information* **26**(6): 383–385.

Senker, P. (1986). *Towards the Automatic Factory.* IFS, Bedford.

Sennett, R. (1998). *The Corrosion of Character: The Personal Consequences of Work in the New Capitalism.* W.W. Norton and Company, New York.

Simon, H. (1962). 'The architecture of complexity'. *Proceedings of the American Philosophical Society* **106**(6): 467–482.

Slaughter, S. (1993). Innovation and learning during implementation: a comparison of user and manufacturer innovations. *Research Policy* **22**(1): 81–95.

Soubra, Y. (Ed.) (1993). *Information Technology and International Competitiveness: The Case of the Construction Services Industry.* United Nations, Geneva.

Stinchcombe, A.L. (1990). *Information and Organisations.* University of California Press, Berkeley.

Sykes, A. (1995). *Product Standards for Internationally Integrated Goods Markets.* The Brookings Institution, Washington, DC.

Tang, P., Curry, R. and Gann, D.M. (2000). *Telecare: New Ideas for Care @ Home.* Policy Press, Bristol.

Teece, D.and Pisano, G. (1994). The dynamic capabilities of firms: an introduction. *Industrial and Corporate Change* **3**(3): 537–556.

Tidd, J., Bessant, J. and Pavitt, K. (1997). *Managing Innovation.* Wiley, Chichester.

Touche Ross (1991). *Office Automation: The Barriers and Opportunities.* Touche Ross and IAM, London.

Towill, D.R. (1997). Successful Business Systems Engineering Parts I and II. *IEE Engineering Management Journal* Vol **7**(Nos. 1 & 2): 55–64 and 89–96.

Turin, D.A. (1966). What do we mean by building? University College, London, p. 22.

Turin, D.A. (1967). Building as a process. *Transactions of the Bartlett Society*: 87–108.

Turin, D.A., (Ed.) (1975). *Aspects of the Economics of Construction.* Godwin, London.

Venturi, R. (1977). *Complexity and Contradiction in Architecture.* Butterworth Architecture, London.

Voeller, J. (1995). *The Likely Progress of Engineering Automation Over the Next Decade.* CII Annual Conference, Austin, Texas.

von Hippel, E. (1988). *The Sources of Innovation.* Oxford University Press, Oxford.

Wachsmann, K. (1961). *The Turning Point of Building.* Reinhold Publishing, New York.

Walsh, V. (1984). Invention and innovation in the chemical industry: demand pull or discovery push'. *Research Policy* **13**(4): 211–234.

Westney, D.E. (1987). *Managing innovation in the information age: the case of the building industry in Japan*. Symposium on Managing Innovation in Large Complex Firms, INSEAD, Fontainebleau.

White, R.B. (1965). *Prefabrication*. HMSO, London.

Winch, G.E. (1999). *Innovation in the British Construction Industry: The Role of Public Policy Instruments*. The UK TG35 Team, Working Paper, London.

Womack, J.P. and Jones, D.T. (1996). *Lean Thinking*. Simon & Schuster, London.

Womack, J.P., Jones, D.T. and Roos, D. (1990). *The Machine that Changed the World*. Maxwell Macmillan International, New York.

Woodward, J. (1965). *Industry and Organisation: Theory and Practice*. Oxford University Press, Oxford.

Wootton Jeffreys Consultants Ltd and Thorpe, B. (1991). *An Examination of the Effects of the Use Classes Order 1987 and the General Development Order 1988*. HMSO, London,.

Wright, R.N., Rosenfeld, A.H. and Fowell, A.J. (1995). *National Planning for Construction and Building R&D*. National Institute of Standards and Technology, Gaithersburg NISTIR 5759.

Yeang, K. (1994). *Bioclimatic Skyscrapers*. Artemis, London.

Zuboff, S. (1988). *In the Age of the Smart Machine: The Future of Work and Power*. Basic Books, New York.

Index